student book volume **2**

Cultural, Social and Technical

VISI**O**ns

MATHEMATICS

Secondary Cycle Two, Year Two

Claude Boivin
Dominique Boivin
Antoine Ledoux
Étienne Meyer
Nathalie Ricard
Vincent Roy

LES ÉDITIONS
CEC
A Quebecor Media Company

9001, boul. Louis-H.-La Fontaine, Anjou (Québec) Canada H1J 2C5
Telephone: 514 351-6010 • Fax: 514 351-3534

ORIGINAL VERSION

Publishing Manager
Véronique Lacroix
Anick L'Écuyer

Production Manager
Danielle Latendresse

Coordination Manager
Rodolphe Courcy

Project Manager
Diane Karneyeff

Proofreader
Viviane Deraspe

Graphic Design
Dessine-moi un mouton

Technical Illustrations
Stéphan Vallières

General Illustrations
Yves Boudreau

Iconographic Research
Jean-François Beaudette

These programs are funded by Quebec's Ministère de l'Éducation, du Loisir et du Sport, through contributions from the Canada-Québec Agreement on Minoroty-Language Education and Second-Language Instruction.

Visions, Cultural, Social and Technical, *Student Book*, Volume 2, Secondary Cycle Two, Year Two
© 2009, Les Éditions CEC inc.
9001, boul. Louis-H.-La Fontaine
Anjou (Québec) H1J 2C5

Translation of *Visions, Culture, société et technique*, volume 2, ISBN 978-2-7617-2595-8
© 2009, Les Éditions CEC inc.

Legal Deposit: 2009
Bibliothèque et Archives nationales du Québec
Library and Archives Canada

ISBN 978-2-7617-2797-6
Printed in Canada
3   4   5   6   7   15   14   13   12   11

The authors and publisher wish to thank the following people for their collaboration in the evolution of this project.

Scientific Consultants
Driss Boukhssmi, Professor, Université du Québec en Abitibi-Témiscamingue
Matthieu Dufour, Professor, Université du Québec à Montréal

Collaboration
Stéphane Brosseau, Teacher, École Secondaire l'Horizon, CS des Affluents
Yanick L'Écuyer, Teacher, Collège Champagneur
Guillaume Marcoux, Teacher, Collège François-de-la-Place
François Maupetit, Teacher, École Secondaire Les Compagnons-de-Cartier, CS des Chênes
Ray Venables, Teacher, Evergreen High School, CS Eastern Shores

ENGLISH VERSION

Translation and Linguistic Review
Donna Aziz
Maki Fukushima
Alain Groven

Pedagogical Review
Joanne Malowany
Pat Ryan, consultant

Proofreader
Aalia Persaud

Project Manager
Patrick Bérubé
Rita De Marco
Valerie Vucko

A special thank you to the following people for their collaboration in the evolution of this project.

Collaboration
Michael J. Canuel
Robert Costain
Margaret Dupuis
Rosie Himo
Doris Kerec
Adam Koutsoukos
Louis-Gilles Lalonde
Denis Montpetit
Bev White

Les Éditions CEC Inc. thank the Government of Québec for the financial assistance granted for the publishing of this book through the Tax credit for book publishing program, administered by SODEC.

TABLE OF CONTENTS

VISI6n

PRESENTATION OF STUDENT BOOK

This *Student Book* contains three chapters each called "Vision." Each "Vision" presents various "Learning and evaluation situations (LES)" sections and special features "Chronicle of the past," "In the workplace" and "Overview." At the end of the *Student Book*, there is a "Reference" section.

REVISION

The "Revision" section helps to reactivate prior knowledge and strategies that will be useful in each "Vision" chapter. This feature contains one or two activities designed to review prior learning, a "Knowledge summary" which provides a summary of the theoretical elements being reviewed and a "Knowledge in action" section consisting of reinforcement exercises on the concepts involved.

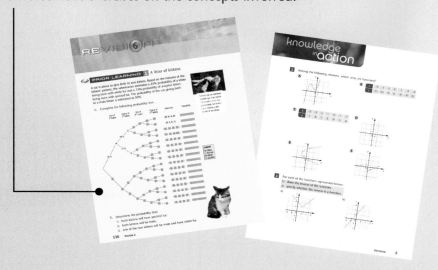

THE SECTIONS

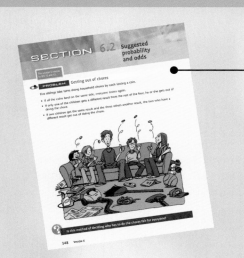

A "Vision" chapter is divided into sections, each starting with a problem and a few activities, followed by the "Technomath," "Knowledge" and "Practice" features. Each section is related to a LES that contributes to the development of subject-specific and cross-curricular competencies, as well as to the integration of mathematical concepts that underscore the development of these competencies.

Problem

The first page of a section presents a problem that serves as a launching point and is made up of a single question. Solving the problem engages several competencies and various strategies while calling upon the mobilization of prior knowledge.

Activity

The activities contribute to the development of subject-specific and cross-curricular competencies, require the use of various strategies, mobilize knowledge and further the understanding of mathematical notions. These activities can take on several forms: questionnaires, material manipulation, construction, games, stories, simulations, historical texts, etc.

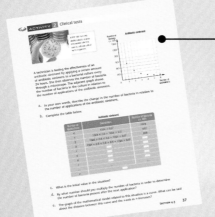

Technomath

The "Technomath" section allows students to use technological tools such as a graphing calculator, dynamic geometry software or a spreadsheet program. In addition, the section shows how to use these tools and offers several questions in direct relation to the mathematical concepts associated with the content of the chapter.

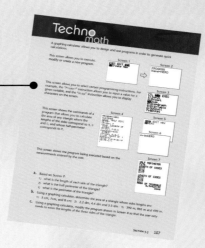

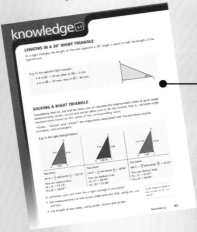

Knowledge

The "Knowledge" section presents a summary of the theoretical elements encountered in the section. Theoretical statements are supported with examples in order to foster students' understanding of the various concepts.

Practice

The "Practice" section presents a series of contextualized exercises and problems that foster the development of the competencies and the consolidation of what has been learned throughout the section.

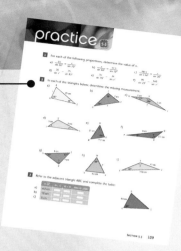

SPECIAL FEATURES

Chronicle of the past

The "Chronicle of the past" feature recalls the history of mathematics and the lives of certain mathematicians who have contributed to the development of mathematical concepts that are directly related to the content of the "Vision" chapter being studied. This feature includes a series of questions that deepen students' understanding of the subject.

In the workplace

The "In the workplace" feature presents a profession or a trade that makes use of the mathematical notions studied in the related "Vision" chapter. This feature includes a series of questions designed to deepen students' understanding of the subject.

Overview

The "Overview" feature concludes each "Vision" chapter and presents a series of contextualized exercises and problems that integrate and consolidate the competencies that have been developed and the mathematical notions studied. This feature ends with a bank of problems, each of which focuses on solving, reasoning or communicating.

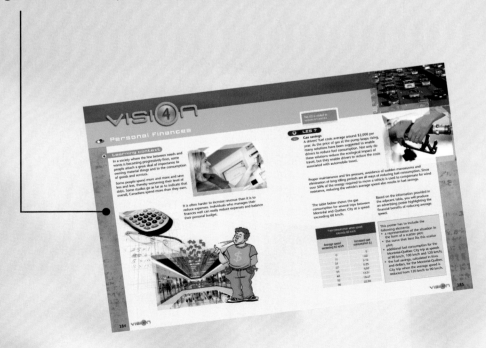

The "Practice" and "Overview" features, include the following:
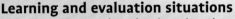
- A number in a blue square refers to a Priority 1 and a number in an orange square a Priority 2.
- When a problem refers to actual facts, a keyword written in red uppercase indicates the subject with which it is associated.

Learning and evaluation situations

The "Learning and evaluation situations" (LES) are grouped according to a common thematic thread; each focuses on a general field of instruction, a subject-specific competency and two cross-curricular competencies. The knowledge acquired through the sections helps to complete the tasks required in the LES.

REFERENCE

Located at the end of the *Student Book*, the "Reference" section contains several tools that support the student-learning process. It consists of two distinct parts.

The "Technology" part provides explanations pertaining to the functions of a graphing calculator, the use of a spreadsheet program as well as the use of dynamic geometry software.

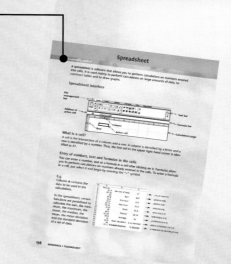

The "Knowledge" part presents notations and symbols used in the *Student Book*. Geometric principles are also listed. This part concludes with a glossary and an index.

ICONS

 Indicates that a worksheet is available in the *Teaching Guide*.

 Indicates that the activity can be performed in teams. Details on this topic are provided in the *Teaching Guide*.

 Indicates that some key features of subject-specific competency 1 are mobilized.

 Indicates that some key features of subject-specific competency 2 are mobilized.

 Indicates that some key features of subject-specific competency 3 are mobilized.

 Indicates that subject-specific competency 1 is being targeted in the LES.

 Indicates that subject-specific competency 2 is being targeted in the LES.

 Indicates that subject-specific competency 3 is being targeted in the LES.

VISI◯N 4

From functions to modelling

How can you choose the telephone plan that best suits your needs? Can one predict how stock values will change? How are income tax rates calculated? Various activities can be represented using mathematical models. What family of functions corresponds to each of these models? In "Vision 4," you will analyze and model various situations and determine the characteristics applicable to each of them. You will use mathematical models to make predictions and informed decisions.

Arithmetic and algebra

- Relations, functions and inverse of functions
- Second-degree polynominal functions
- Exponential functions
- Periodic functions
- Step functions
- Piecewise functions
- Modelling a situation

Geometry   **Statistics**   **Probability**

RE VISI④N

| PRIOR LEARNING | 1 | Cardiofrequency meters |

A cardiofrequency meter is a device that allows athletes to measure their heart rate during a workout or sports events. A strap worn around the chest monitors the heartbeat and sends this information to a device that displays and records data such as maximum, minimum and average heart rates. Some cardiofrequency meters include Global-Positioning-System (GPS) technology that measures distances covered by an athlete.

The table below contains data recorded by Matteo's cardiofrequency meter during his last two workout sessions.

Tuesday's workout

Time (min)	Heart rate (beats/min)
2	80
4	112
6	120
8	124
10	120
12	116
14	124
16	140
18	120
20	116

Thursday's workout

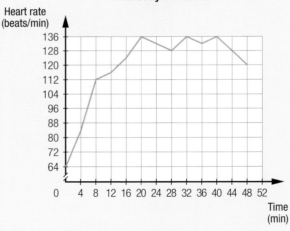

a. Create a scatter plot of the data collected during Tuesday's workout.

b. Complete the following table.

Workout session	Maximum heart rate	Minimum heart rate	Periods when heart rate increased	Periods when heart rate decreased
Thursday				

c. What was Matteo's heart rate 24 min after the start of Thursday's workout?

Self-employed workers

Self-employed workers perform work or provide goods and services to clients who, in turn, pay these workers on the basis of a predetermined price. Self-employed workers cover their own expenses and are responsible for the financial risks associated with their work.

The graph below represents a self-employed worker's financial status over the first 12 months of business activities.

Financial status

Assets ($)

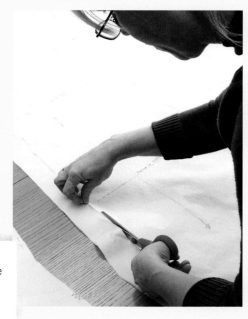

Self-employed workers may represent various sectors of the workforce such as accounting, fashion design, graphic design and journalism.

a. What is the *y*-intercept of this function, and what does it represent in relation to the context?

b. What is the *x*-intercept of this function, and what does it represent in relation to the context?

c. What is the sign of this function, and what does it represent in relation to the context?

d. Is the value of this self-employed worker's assets increasing or decreasing? Explain your answer.

e. Determine the domain and the range of this function.

f. Is the inverse of the function that represents this self-employed worker's financial status also a function? Explain your answer.

g. Determine the domain and the range of the inverse of this function.

RELATIONS, INDEPENDENT VARIABLES AND DEPENDENT VARIABLES

The link between two variables is called a **relation**.

In general, in a relation between two variables the following is true:

- The variable whose variation **generates** the other's variation is called an **independent variable**.
- The variable whose variation **reacts** to the other's variation and is called a **dependent variable**.

E.g.

Relation	Independent variable	Dependent variable
1) The mass and cost of a frozen turkey	Mass ⟶ Cost The cost of a frozen turkey depends on its mass.	
2) The total area of the walls and ceiling of a room and the time required to paint this room	Total area ⟶ Time The time required to paint a room depends on the total area of the walls and ceiling.	

INVERSE

An inverse relation, or simply the **inverse**, is obtained by exchanging the values of each ordered pair in a relation between two variables.

E.g.

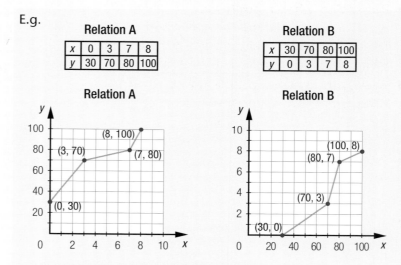

Relation A

x	0	3	7	8
y	30	70	80	100

Relation B

x	30	70	80	100
y	0	3	7	8

Relation **B** is the inverse of Relation **A** and vice versa.

FUNCTION

A relation between two variables is called functional, or simply a **function**, when no more than one value of the dependent variable is associated with each value of the independent variable.

In the graphical representation of a function, no more than one y-coordinate is associated with each x-coordinate.

E.g.

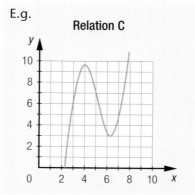

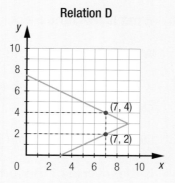

Relation **C** is a function. Relation **D** is not a function since the x-coordinate 7 is associated with more than one y-coordinate, in this case 2 and 4.

PROPERTIES OF FUNCTIONS

Domain and range (image)

The **domain** of a function is the set of all the possible values of the **independent variable**.

The **range** or **image** of a function is the set of all the possible values of the **dependent variable**.

E.g.

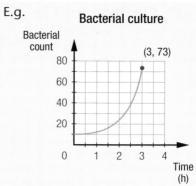

Bacterial culture

Domain: [0, 3] h
Range: [10, 11, 12, ..., 73] bacteria

Variation: increase, decrease and constant

Over an interval of the domain, a function is:

- **increasing** when a positive or negative variation of the independent variable generates, respectively, a positive or negative variation of the dependent variable

- **decreasing** when a positive or negative variation of the independent variable generates, respectively, a negative or positive variation of the dependent variable

- **constant** when a variation of the independent variable does not generate any variation of the dependent variable

E.g.

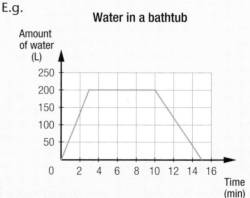

Water in a bathtub

Increase: [0, 10] min
Constant: [3, 10] min
Decrease: [3, 15] min

Extrema: minimum and maximum

The **minimum** of a function is the smallest value of the dependent variable.

The **maximum** of a function is the largest value of the dependent variable.

E.g.

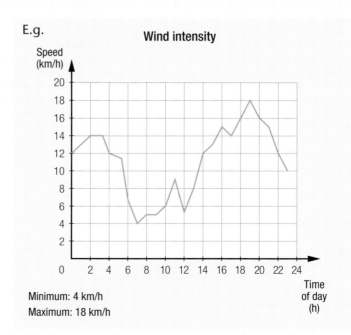

Wind intensity

Minimum: 4 km/h
Maximum: 18 km/h

The sign: positive or negative

Over an interval of the domain, a function is:

- **positive** if the values of the dependent variable are positive
- **negative** if the values of the dependent variable are negative

E.g.

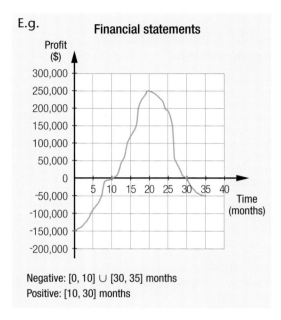

Negative: [0, 10] ∪ [30, 35] months
Positive: [10, 30] months

Intercept coordinates: *x*-intercept (zero) and *y*-intercept (initial value)

The **zero of a function** is the value of the independent variable when that of the dependent variable is zero. Graphically, the zero corresponds to the *x*-intercept, which is the *x*-coordinate of the intersection point of the curve and the *x*-axis.

The **initial value of a function** is the value of the dependent variable when that of the independent variable is zero. Graphically, the initial value is the **y-intercept**, meaning the *y*-coordinate of the intersection point of the curve and the *y*-axis.

E.g.

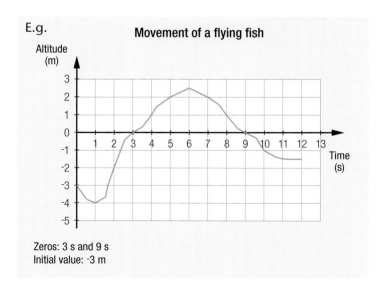

Zeros: 3 s and 9 s
Initial value: -3 m

knowledge in action

1 Among the following relations, which ones are functions?

A

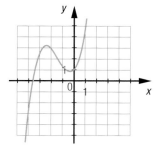

B

x	-3	-2	-1	0	1	2	3
y	-27	-12	-3	0	-3	-12	-27

C

x	-3	-2	-1	-1	0	1	2
y	1	6	7	8	9	3	4

D

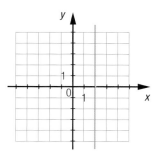

E

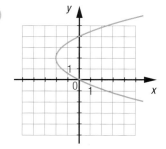

F

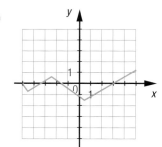

2 For each of the functions represented below, do the following:

1) Draw the inverse of the function.

2) Specify whether the inverse is a function.

a)

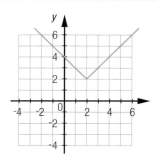

b)

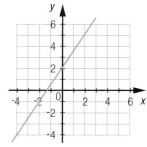

3 For each of the functions represented below, determine:

1) the domain and the range
2) the initial value
3) the extrema
4) the variation
5) the zero(s)
6) the sign

a)

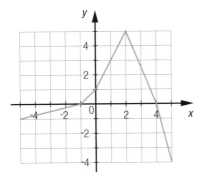

b)

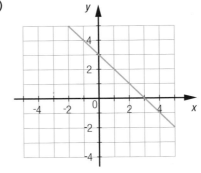

c)

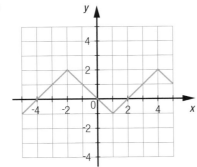

d)

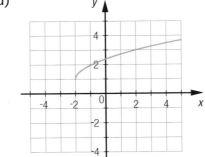

4 A taxi company's rates are as follows: $3.15 at the time of pick-up, meaning as soon as the customer enters the taxi, and $1.30 for each kilometre travelled.

a) Complete the table below.

Cost of a taxi ride

Distance covered (km)	Cost ($)
0	
5	
10	
15	
20	
25	
30	

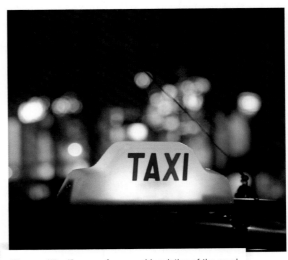

b) What would the cost be for a 17-km ride?

c) If a taxi ride costs $33.05, what distance was travelled?

The word "taxi" comes from an abbreviation of the word "taximeter" which is the counter that sets the fare of a taxi ride in relation to the distance travelled.

5 The graph below provides information concerning a subway station's passenger traffic over the course of the day.

Ile Notre Dame and Ile Sainte-Hélène together form Parc Jean-Drapeau. Ile Notre Dame was built with earth and rock from the excavation of the Montréal subway. It took 28 million tons of earth and rock to build the island!

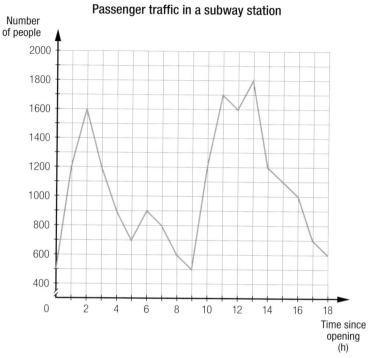

Passenger traffic in a subway station

a) During the day, when is the passenger traffic level at a:
   1) maximum?
   2) minimum?

b) When is the passenger traffic at 1200?

c) During which time intervals is the passenger traffic more than 1200 people?

d) Determine the variation:
   1) over the interval [0, 2] h
   2) over the interval [2, 9] h

6 A function *f* has the following properties.

- Domain: [-5, 5]
- Range: [-6, 5]
- Initial value: {0}
- Zeros: {-5, 0, 2}
- The function is increasing over [-5, -2] ∪ [1, 4].
- The function is decreasing over [-3, 1] ∪ [4, 5].
- The function is positive over [-5, 0] ∪ [2, 5].
- The function is negative over [0, 2].
- Maximum: 5
- Minimum: -6

On a Cartesian plane, draw a curve that can be associated with this function.

7 A landscaping company was hired to lay down the sod on the soccer and football fields of a municipal park. It would take 720 h for a single employee to perform the work.

a) Complete the adjacent table of values.

b) Represent this situation graphically.

c) Is the function associated with this situation increasing or decreasing? Explain your answer.

d) Does this situation represent a proportional situation? Explain your answer.

Time required to lay down the sod	
Number of employees	Time (h)
1	
2	
3	
4	
5	
6	
8	
10	
12	

"Sod" is a slab of grass. Sod is an organic product of partially decayed vegetation matter that consists mainly of moss and peat and is used most often in horticulture as well as in industry. Canada is one of the most important sod producers in the world.

8 Consider the following graphical representations of two functions:

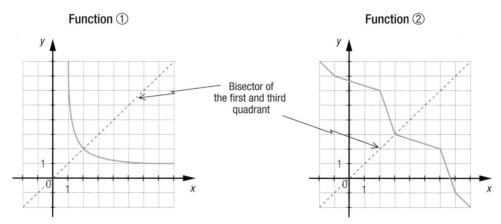

Function ①   Function ②

Bisector of the first and third quadrant

a) State one common feature of the graphical representations of the two functions shown above.

b) Draw the inverse of each of these functions. What do you notice?

c) What conjecture can you formulate regarding the inverse of functions exhibiting the feature found in **a)**?

This section is related to LES 7, 8 and 9.

PROBLEM Spring floods

In Québec, the spring thaw causes the water levels in most rivers to rise. Depending on the amount of snow that accumulates and how quickly it melts, this rise can increase slowly or quickly. The rising water levels can sometimes pose a threat to riverside residents.

The following table contains information from three observation points on the Rivière aux Mûres:

Rise in Rivière aux Mûres water levels

Time elapsed from the start of the flood (h)	Water level at observation point A (mm)	Water level at observation point B (mm)	Water level at observation point C (mm)
0	150	110	125
1	150.5	115.5	123
2	151	112.5	128
3	151.5	118	126
4	152	115	131
5	152.5	120.5	129

If the water level reaches 160 mm at any of the observation points, an evacuation of the riverside residents will be initiated.

If the water level continues to rise at the same rate, when will evacuation procedures be initiated?

ACTIVITY 1 Earnings and expenses

According to Québec labour standards, employers have one month to pay new employees. The employee's pay must thereafter be given at regular intervals not exceeding 16 days or one month in the case of executives or contract employees.

The following graphs represent the financial situation of four plant employees; each earns a weekly salary of $600.

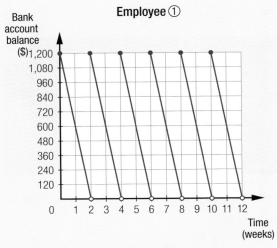

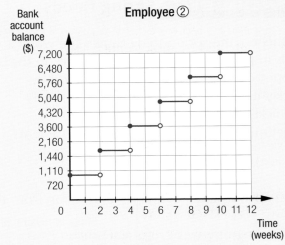

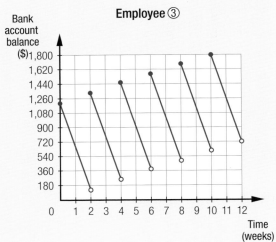

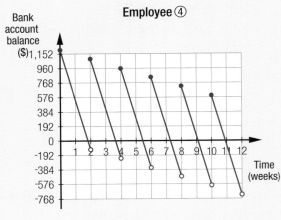

a. Based on the financial situation of Employee ①:

1) what is the maximum balance in his bank account?

2) at what time(s) does the employee's financial situation indicate assets of $600?

b. Based on the financial situation of Employee ②:

1) what is the initial amount in his bank account?

2) what does the length of each segment on the graph represent?

c. In your own words, describe each employee's financial situation.

d. If the trend of the financial situation of each employee continues, what will the balance be in their account by week 14?

ACTIVITY 2 P/BiT

P/BiT substances, pronounced "pee-bit," are listed by Environment Canada as being:

- Persistent (P): They are resistant to natural biological break down.

- Bioaccumulative (B): They are stored and accumulated in living tissue and are transferred up the food chain.

- Inherently toxic (iT): They are proven to be harmful to human health or to the environment.

Dioxins are one type of P/BiT. Once they have entered the body, these substances cannot be naturally eliminated except by transfer through placenta or breast milk. The people most seriously affected by dioxins are babies who may suffer damage to their immune, endocrine and nervous systems.

The scatter plot below shows the evolution, from 1976 to 2008, of P/BiT concentration in breast milk among Swedish women.

a. Which of these curves best fits the scatter plot?

b. Associate each curve with one of the of functions listed below:

1) first-degree polynomial function whose rule is $y = 0.14x - 0.56$

2) second-degree polynomial function whose rule is $y = \dfrac{x^2}{200}$

3) exponential function whose rule is $y = 0.24 \, (1.1)^x$

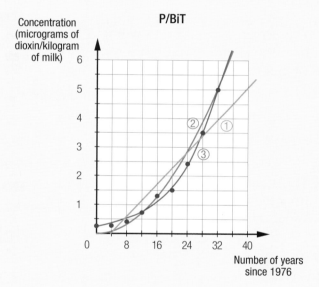

c. What type of function could serve as the best mathematical model to describe this situation?

d. Based on the information provided, what will the concentration of dioxin in breast milk be by 2012?

Health Canada recognizes that breast milk is the best way to feed newborn babies. Dioxin concentrations found in breast milk in Canadian women, which were very low to begin with, decreased 50% between 1980 and 1990. Dioxin concentrations released in the environment, on the other hand, have decreased 60% since 1990.

Techno math

A graphing calculator allows you to display scatter plots, detect graphical trends and deduce which mathematical models are best suited to represent various situations.

This table of values displays the data collected in the course of an experiment in which variables y_1 and y_2 vary in relation to variable x.

x	y^1	y^2
1	3	42
2	5	11
3	9	40
4	16	10
5	28	39
6	45	12

This screen allows you to enter and edit experimental data.

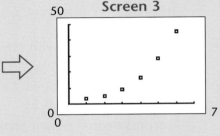

Screen 1

This screen allows you to define how the relationship between variables x and y_1 is displayed.

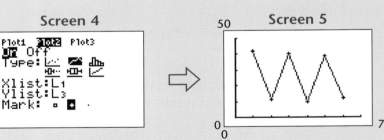

Screen 2

Screen 3

This screen allows you to define how the relationship between variables x and y_2 is displayed.

Screen 4

Screen 5

a. What would be the graphical effect of selecting option Off rather than option On on Screen **2** or Screen **4**?

b. List the differences between Screens **2** and **4**.

c. Of Screen **3** and Screen **5**, which one contains data that could be graphically modelled with:
1) a curve with a recurring pattern?
2) a curve which, from left to right, gets steeper more and more quickly?

d. Using a graphing calculator, display a scatter plot comprised of six points that can be graphically modelled:
1) with a straight line
2) with a curve which from left to right gets steeper less and less quickly
3) with a horizontal line
4) with two lines that have different slopes

FAMILIES OF FUNCTIONS

Depending on the link between two variables, it is possible to represent different situations of daily life with **mathematical models**, in other words, with functions whose behaviour is both known and predictable. More specifically, these models allow you to analyze a situation or to make certain predictions. Below are several functions that could serve as mathematical models for a given situation:

Zero-degree polynomial function

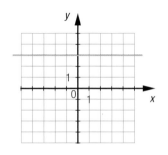

First-degree polynomial function

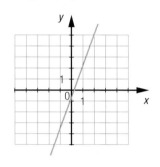

Second-degree polynomial function

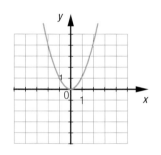

Inverse variation function

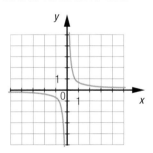

Exponential function

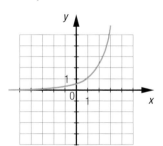

Step function

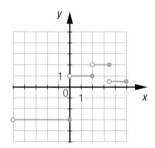

Periodic function

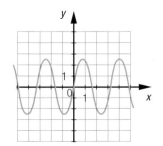

Piecewise function

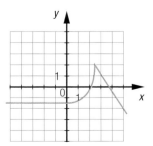

CHOOSING A MODEL

The information in a table of values or a scatter plot associated with a situation involving two variables does not always exhibit a systematic regularity or result in points arranged according to a perfectly defined trend. This may be due to errors in manipulation, measurement or to the degree of precision of the instrument used. By analyzing the shape of the scatter plot or by comparing certain properties of functions, one can choose a mathematical model that makes it possible to visualize a situation.

E.g. 1) The information below represents the concentration of greenhouse gases in the atmosphere from 1950 to 2000:

Greenhouse gases

Year	1950	1960	1970	1980	1990	2000
Concentration of CO_2 (ppm)	50	80	100	300	800	1800

The scatter plot representing this situation shows a trend associated with an exponential function, in other words, a function with an increasingly steep curve.

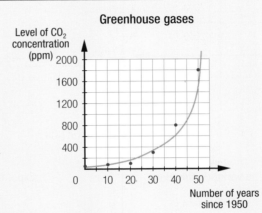

2) Consider the time required to assemble a stage in relation to the number of employees assigned to this task.

Building a stage

Number of employees	2	4	6	8	10	12	14	16	18	20
Time (h)	22	12	7	6	4.5	4	3	3	2.5	2.25

The scatter plot representing this situation exhibits a trend associated with an inverse variation function, since the graphical representation shows that the ends of the curve approach the axes more and more slowly without ever touching them.

1 Determine the type of function for each of the following situations.

a)
Frog population at Blueberry Pond
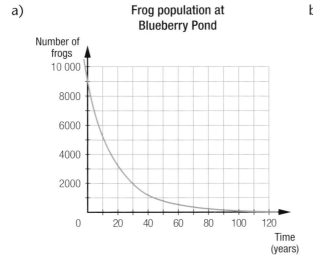

b)
Amount of gas left in a car's gas tank
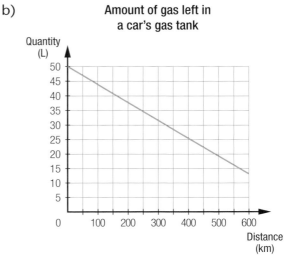

c)
Electrocardiogram
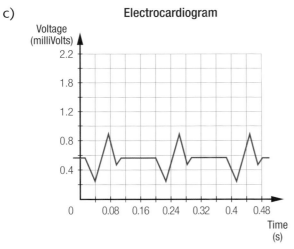

d)
Alternating current
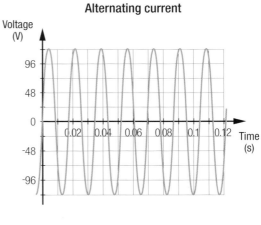

e)
Snow accumulation
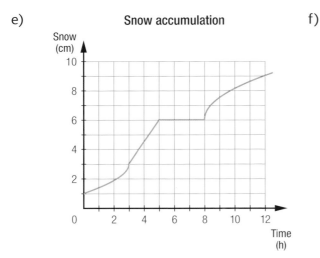

f)
Cottage rental rates
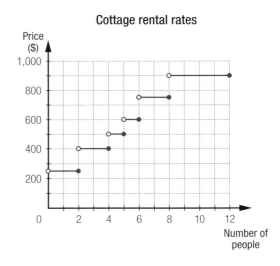

2 Match each situation in the left column with the a function in the right column.

Situation	Function
A A person pays $68.25 for her unlimited monthly bus pass.	**1** Zero-degree polynomial function
B Studying the phases of the Moon.	**2** First-degree polynomial function
C The area of a circle is equal to the product of π and the length of its radius squared.	**3** Second-degree polynomial function
D The thickness of a book in relation to the number of pages it contains.	**4** Exponential function
E The number of cells of an embryo doubles every 18 h.	**5** Periodic function

 3 For each of the graphs below, do the following:

a) Determine what type of function best represents the scatter plot.

b) Draw a curve that best represents the data shown on the graph.

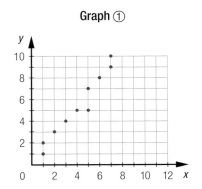

Graph ①

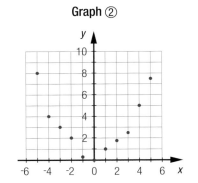

Graph ②

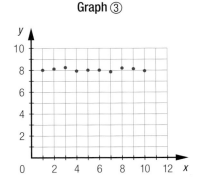

Graph ③

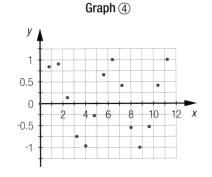

Graph ④

4 For each of the tables of values below, determine:

    a) the type of function represented   b) the rule of the function represented

    c) the value of y when $x = 12.5$

Table of values

x	y
-7	6
-5	6
-3	6
-1	6
1	6
3	6
5	6
7	6

Table of values

x	y
1	120
2	60
3	40
4	30
5	24
6	20
8	15
10	12

Table of values

x	y
0	0
1	3
2	6
3	9
4	12
5	15
6	18
7	21

Table of values

x	y
-3	9.5
-2	9
-1	8.5
0	8
1	7.5
2	7
3	6.5
4	6

5 a) For each of the graphs below, name the type of function represented.

Function ①

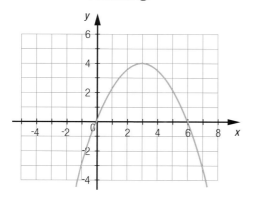

Function ②

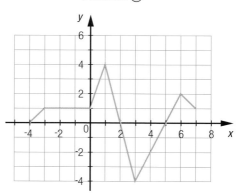

Function ③

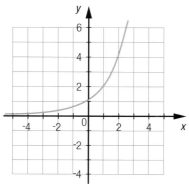

Function ④

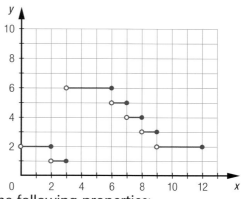

    b) For each of the functions above, determine the following properties:

      1) the domain and the range   2) the variation

      3) the sign   4) the extrema

      5) the initial value   6) the zero(s)

6 The adjacent graph displays the outside temperature during the first 12 hours of a fall day.

a) What type of function is represented in the adjacent graph?

b) What is the initial temperature?

c) What is the lowest temperature?

d) At what time was the recorded temperature:

  1) lowest?   2) highest?

e) At what time(s) is the temperature:

  1) 0°C?   2) negative?   3) positive?

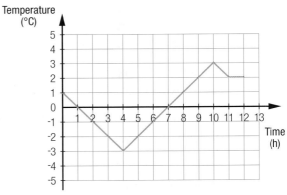

Outside temperature

7 The adjacent table indicates the fines for speeding on a road where the speed limit is 100 km/h.

Speeding fines

Speed (km/h)	Fine ($)
[0, 100]	0
]100, 130]	95
]130, 140]	220
Over 140	295

a) Complete the table of values below.

Fines for speeding

Speed (km/h)	100	105	110	115	120	122	124	126	128	130	132	134	136	138	140
Fine ($)															

b) Draw a graphical representation of the data obtained in **a)**.

c) What type of function should serve as a mathematical model for this situation?

The maximum speed limit on Québec highways is 100 km/h. This limit is set in relation to ideal conditions and drivers. One must slow down when it snows or rains. Each year, approximately 200 people lose their lives in accidents caused by speeding. Next to drinking and driving, speeding is the second cause of highway deaths.

8 As shown in the adjacent graph, a restaurant owner used the second-degree polynomial function to model his dinner patronage. Based on this model, answer the following:

a) What is the maximum number of patrons in this restaurant at suppertime?

b) For how many hours does the number of patrons increase?

c) How many patrons are there in the restaurant at 6:00 p.m.?

d) At what times are there 25 patrons in the restaurant?

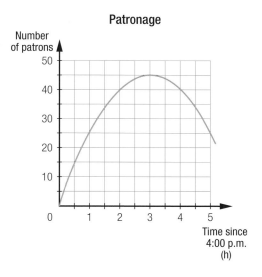

Patronage

Average Canadians spend one-third of their food budget in restaurants. The most popular dish is fries, and the most popular beverage is coffee.

9 Consider the graphical representation below that represents an experiment in a water basin:

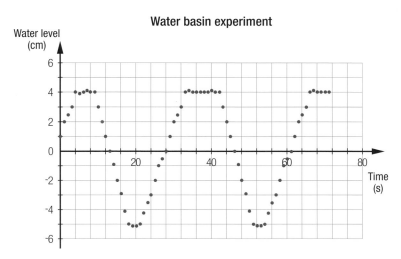

Water basin experiment

a) Determine the type of function best suited to represent the data whose x-coordinate is within the interval:

1) [0, 4] s   2) [9, 18] s

3) [4, 9] s   4) [0, 18] s

b) What type of function should serve as a mathematical model for the data collected by this experiment?

SECTION 4.2 Second-degree polynomial function

This section is related to LES 7.

 PROBLEM Recording space

If you observe a compact disc (CD), you can see the distinction between the recorded space and the unused disc space.

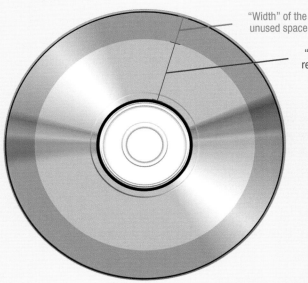

"Width" of the unused space

"Width" of the recorded space

The table below shows the relation between the "width" of the recorded space and the length of the recording.

A compact disc is a circular plate made of plastic and aluminum. There is a continuous spiral track that runs from the centre to the edge of the disc. Discs are manufactured using lasers that etch a series of pits. The zones between pits are called "lands." Change from pits to lands and vice versa forms a binary code that is read by the CD reader and is transformed into music or software.

Compact disc

Width of recorded space (mm)	Duration of recording
0	0 min 0 s
5	1 min 36 s
10	6 min 24 s
15	14 min 24 s
20	25 min 36 s
25	40 min 0 s

Can 80 min of music be recorded on a CD where the "width" of the recordable surface measures 3 cm?

ACTIVITY 1 Free fall

On August 16, 1960, Joseph Kittinger took part in an aeronautical medical research project during which he flew in an open gondola carried aloft by a large helium balloon up to an altitude of 31 300 m. Wearing a pressurized suit, Kittinger jumped overboard in order to test the effects of high altitude ejection on the human body.

Joseph Kittinger, born in Florida in 1928, was a US Air Force pilot. He established the world record for the highest parachute jump.

The adjacent table of values provides a description of Kittinger's fall for the first few seconds before deployment of his main parachute.

a. Is the change in the vertical distance best represented by:
   1) a zero-degree function? Explain your answer.
   2) a first-degree function? Explain your answer.
   3) an inverse variation function? Explain your answer.

b. Complete the variations in the adjacent table of values.

c. The variations obtained in **b)** are of what type?

d. Complete Column **B** in the table below.

Free fall

Time (s)	Vertical distance covered (m)
0	0
24	195.84
48	783.36
72	1762.56
96	3133.44
120	4896
144	7050.24
168	9596.16
192	12 533.76
216	15 863.04
240	19 584
264	23 696.94

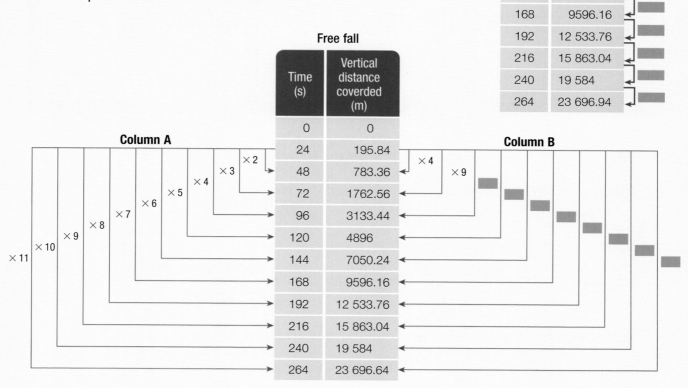

Free fall

Time (s)	Vertical distance coverded (m)
0	0
24	195.84
48	783.36
72	1762.56
96	3133.44
120	4896
144	7050.24
168	9596.16
192	12 533.76
216	15 863.04
240	19 584
264	23 696.64

e. What relationship can you establish between the numbers in Column **A** and Column **B**?

f. What distance did Kittinger vertically travel after 288 s?

ACTIVITY 2 Stretching, compressing and reflecting

Below are several polynomial curves drawn with a graphing tool.

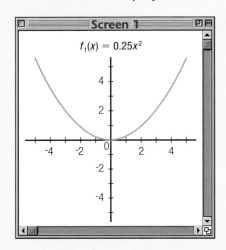

Screen 1
$f_1(x) = 0.25x^2$

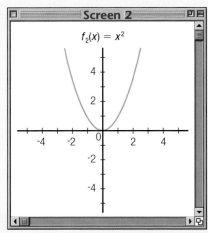

Screen 2
$f_2(x) = x^2$

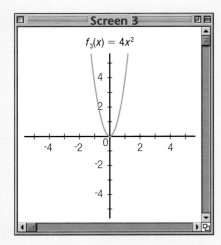

Screen 3
$f_3(x) = 4x^2$

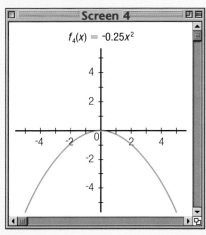

Screen 4
$f_4(x) = -0.25x^2$

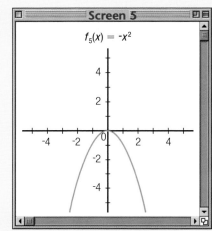

Screen 5
$f_5(x) = -x^2$

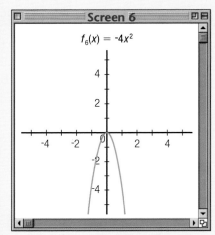

Screen 6
$f_6(x) = -4x^2$

a. What type of function is diplayed in the screens above?

b. Comparing Screens **1** to **3**, what difference can you see:
   1) between the rules of the functions?   2) between the curves?

c. What conjecture can you formulate from the answers found in **b.**?

d. Comparing Screens **4** to **6**, what difference can you see:
   1) between the rules of the functions?   2) between the curves?

e. What conjecture can you formulate from the answers found in **d.**?

f. What difference do you notice in each screen when you compare it to the one situated below it:
   1) between the rules of the functions?   2) between the curves?

g. What conjecture can you formulate from the answers found in **f.**?

Techno math

A graphing calculator allows you to simultaneously display the curves of several functions on the same Cartesian plane. Below is an exploration of some of the graphical features of a function whose rule is expressed in the form $y = ax^2$.

The adjacent screens display the rules for three second-degree polynomial functions as well as their graphical representations.

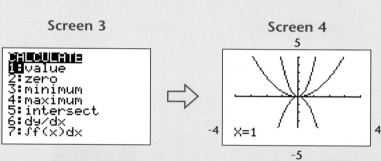

Screen 1   Screen 2

This screen displays various calculations that can be performed on the display screen.

Screen 3   Screen 4

It is possible to view the coordinates of a point on a curve for a chosen x-value.

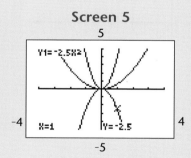
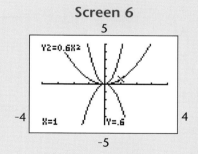
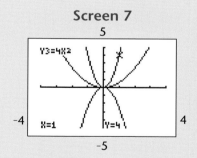

Screen 5   Screen 6   Screen 7

a. What do the curves on Screen **2** have in common?

b. Determine the coefficient of the x^2 for each of the rules on Screen **1**.

c. What conclusion can you draw when you compare the rules on Screen **1** and the coordinates of the points on Screens **5**, **6** and **7**?

d. Using a graphing calculator, determine the effect produced on the graph of a function whose rule is expressed in the form $y = ax^2$ when the value of **a**:

1) is positive

2) is negative

3) is closer and closer to 0

4) is further and further away from 0

SECOND-DEGREE POLYNOMIAL FUNCTION

A function whose rule is written as a second-degree polynomial is called a **second-degree polynomial function** or a **quadratic function.**

For a second-degree polynomial whose rule is written as $f(x) = ax^2$ where a ≠ 0, the following is true:

- The dependent variables are proportional to the squares of the independent variables.
- For every unit increase of the independent variable, the variation of the dependent variable forms an arithmetic sequence.
- The graphical representation is a curve, called a **parabola**, which passes through the origin of the Cartesian plane and is symmetrical over the y-axis.
- The intersection of the curve and the axis of symmetry is called the **vertex.**

E.g.

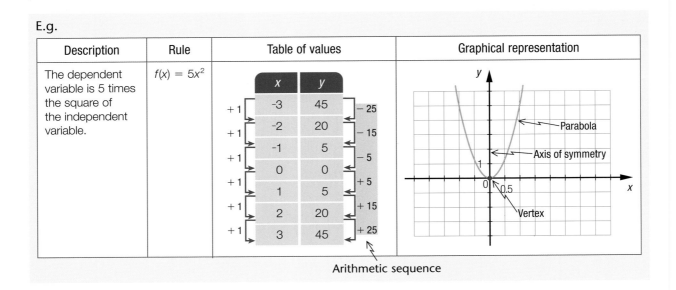

Description	Rule	Table of values	Graphical representation
The dependent variable is 5 times the square of the independent variable.	$f(x) = 5x^2$		

Arithmetic sequence

PARAMETER a

In the rule $f(x) = ax^2$, changing **parameter a** generates a vertical stretch or vertical compression of the graphical representation.

- The further the value of **a** is from 0, the more the curve is vertically stretched.
- The more the value of **a** approaches 0, the more the curve is vertically compressed.
- When the sign of **a** changes, the curve is reflected over the x-axis.

E.g. 1)

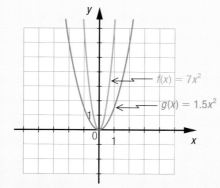

2)

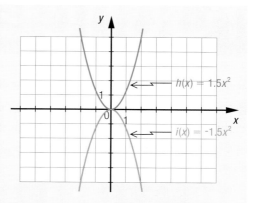

Since 7 is further away from 0 than 1.5, the curve represented by function *f* is more vertically stretched than the curve represented by function *g*.

Since 1.5 is closer to 0 than 7, the curve represented by function *g* is more vertically compressed than the curve represented by function *f*.

Since -1.5 is opposite to 1.5, the curve of function *f* is a reflection of function *g* over the *x*-axis.

DETERMINE THE RULE OF THE SECOND-DEGREE POLYNOMIAL FUNCTION

The rule of a second-degree polynomial function expressed in the form $f(x) = ax^2$ can be determined by following the procedure below.

1. Identify the coordinates of a point on the curve that is not the vertex.	E.g. 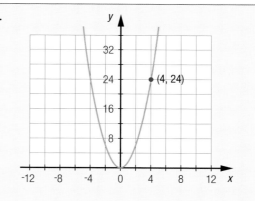
2. Subsitute the *x* and *y* values identified on the curve into the rule of the form $f(x) = ax^2$.	The curve passes through point (4, 24). $24 = a \times 4^2$
3. To find the value of the parameter **a**, solve the resulting equation.	$24 = a \times 4^2$ $24 = a \times 16$ $1.5 = a$
4. Write the rule for the function.	$f(x) = 1.5x^2$
5. Validate the solution.	$24 = 1.5 \times 4^2$ $24 = 1.5 \times 16$ $24 = 24$

1 For each of the following second-degree polynomial functions, do the following:

1) Complete the table of values.
2) Represent the function on a Cartesian plane.

a) $f(x) = 4x^2$

x	y
-6	
-4	
-2	
0	
2	
4	
6	

b) $g(x) = -3x^2$

x	y
-6	
-4	
-2	
0	
2	
4	
6	

c) $h(x) = \frac{1}{6}x^2$

x	y
-6	
-4	
-2	
0	
2	
4	
6	

d) $i(x) = -0.7x^2$

x	y
-6	
-4	
-2	
0	
2	
4	
6	

2 Consider the table of values of the function defined by the rule $f(x) = 7x^2$.

a) Construct a table of values associated with the inverse of function f.

b) Draw the graph of the inverse of function f.

c) Is the inverse of function f a function? Explain your answer.

x	y
-3	63
-2	28
-1	7
0	0
1	7
2	28
3	63

3 Consider the four quadratic functions provided.

Based on the graphical representation of these functions, determine:

a) which is the most vertically stretched

b) which is the most vertically compressed

c) which open upwards

d) which open downwards

$f(x) = 2.5x^2$

$g(x) = 7x^2$

$h(x) = -0.5x^2$

$i(x) = -3.5x^2$

4 a) Complete the adjacent table.

b) What conjecture can you formulate regarding the value of parameter **a** and the extrema of a function whose rule is in the form of $y = ax^2$?

Rule	Coordinates of the vertex	Maximum	Minimum
$f(x) = 16x^2$			
$g(x) = -5x^2$			
$h(x) = 9x^2$			
$i(x) = -3x^2$			

5 Match each of the following rules with the appropriate graph.

A $f(x) = 6x^2$ **B** $g(x) = 12x^2$ **C** $h(x) = -5x^2$ **D** $i(x) = 0.5x^2$

1
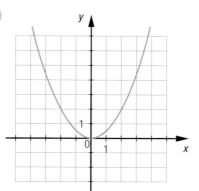

2
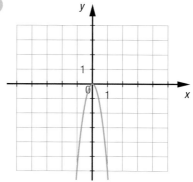

3
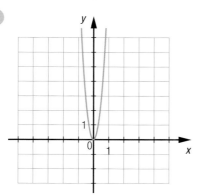

4
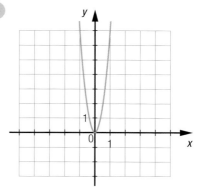

6 The adjacent table of values represent situations related to the area of certain geometric figures.

a) Determine the rule that allows you to calculate area A:
   1) of a square according to the length s of one of its sides
   2) of an isosceles right triangle according to the length l of one of its legs

b) What is the area:
   1) of a square with 6.4 cm sides?
   2) of an isosceles right triangle in which one of the legs measures 10.7 cm?

c) What is the length:
   1) of the side of a square whose area is 324 cm²?
   2) of one of the legs of an isosceles right triangle whose area is 44.18 cm²?

Square		Isosceles right triangle	
Length of side (cm)	Area (cm²)	Length of a leg (cm)	Area (cm²)
1	1	1	0.5
2	4	2	2
3	9	3	4.5
4	16	4	8
5	25	5	12.5

7 Complete the following table.

Rule	Domain	Range	Zero	Increasing interval	Decreasing interval
$f(x) = x^2$					
$g(x) = 6x^2$					
$h(x) = -2x^2$					
$i(x) = -1.5x^2$					

8 The following information pertains to the distance travelled by a remote-controlled car.

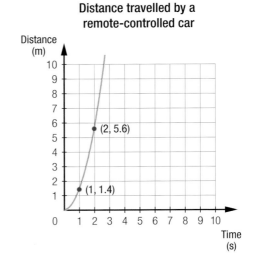

Speed of a remote-controlled car

Speed (m/s) / Time (s)

(3, 8.4)
(3, 5.6)
(1, 2.8)

Distance travelled by a remote-controlled car

Distance (m) / Time (s)

(2, 5.6)
(1, 1.4)

a) Identify the type of function associated with each of the graphs shown above.

b) After 25 s:

   1) what will the speed of the car be?

   2) what distance will the car have travelled?

c) After how many seconds will:

   1) the speed of the car reach 11.5 m/s?

   2) the distance travelled reach 1260 m?

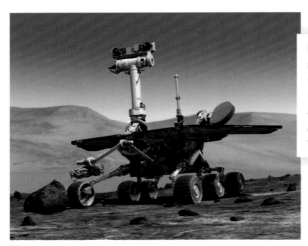

Remote-controlled devices are frequently used in science such as in the Spirit and Opportunity robots sent to Mars in 2004. Their mission was to look for traces of water. The rovers took pictures and gathered geological data and soil information. All this information will help scientists decide whether there ever was life on Mars.

9 Considering that the surface area A of a steel ball with a radius r is calculated using the formula $A = 4\pi r^2$, do the following:

a) Complete the adjacent table of values.

b) Determine the surface area of a steel ball with a radius of 10.02 mm.

Steel balls

Radius (mm)	Surface area (mm²)
1	
2	
3	
4	
5	

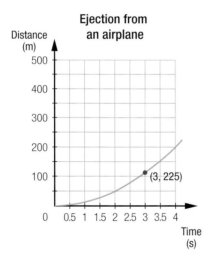

To make steel balls, an amount of molten steel is dropped into a tube. A ball is formed when the molten metal cools.

10 **EJECTION SEAT** During ejection, military aircraft pilots can be subjected to forces 5 to 20 times greater than the force of gravity. The graph below shows the relationship between the time and the distance travelled by a pilot in relation to his ejection point.

Ejection from an airplane

Distance (m) / Time (s)

(3, 225)

Most military aircraft are equipped with ejection seats which allow a pilot to eject when a plane is in danger of crashing. Seconds after ejection, a parachute is deployed, and pilots can land safely.

a) At what distance is the pilot 2.12 s after ejection?

b) How much time after ejection is the pilot 250 m away from the initial point of ejection?

c) The pilot is considered safe if he is more than 300 m from his ejection point. Determine whether or not the pilot is safe if the plane explodes 3.55 s after he has ejected.

11 A flowerbed with a length twice its width is being prepared.

a) What rule allows you to calculate the area A of the flowerbed if its width is represented by x?

b) Draw a graph representing the function whose rule was found in **a)**.

c) How would the curve drawn in **b)** change if the length was equal to:

1) three times the width?   2) the width?

12 **BRAKING DISTANCE** The faster a car moves, the greater the braking distance required. A driver's response, vehicle mass, road conditions and brake and tire wear are all factors that have a bearing on braking distance. The table below provides some braking distances in relation to a vehicle's speed.

Braking distance

Speed (km/h)	Distance (m)
0	0
30	15.5
50	31.2
70	51
100	88.6
120	119

Since 2008, Québec cars must be equipped with winter tires during the winter season. These tires are made of a type of rubber that retains its elasticity down to temperature of -40° C. Such tires can reduce braking distance by 25%.

a) Using a scatter plot, graph this situation.

b) Draw a curve that best represents these points.

c) Find the rule for the mathematical model associated with the curve obtained in **b)**.

d) Determine the braking distance required for a car travelling at:

1) 90 km/h   2) 140 km/h   3) 160 km/h

SECTION 4.3 Exponential functions

This section is related to LES 8.

PROBLEM Achilles and the turtle

Zeno of Elea is a Greek philosopher who wrote a number of paradoxes. One of the best known among Zeno's paradoxes is that of Achilles and the Turtle.

Achilles, who runs 10 times faster than the turtle, gives the latter a 100-m head start in a race. Zeno uses a table similar to the one below to show that Achilles could never have overtaken the turtle.

Zeno of Elea
(ca. 490 - ca. 430 BCE)

Race

Distance covered (m)	
Achilles	Turtle
0	100
100	110
110	111
111	111.1
111.1	111.11
111.11	111.111
111.111	111.1111
…	…

Since I'm 10 times faster than you, I'll give you a 100-m head start.

Why is this situation considered a paradox?

ACTIVITY 1 Worst-kept secret

In the situation illustrated above, each person who is told the secret tells another person a minute later, so the total number of people who know the secret doubles every minute.

a. Complete the adjacent table.

b. Draw a graphical representation of this situation.

c. What type of function best represents this situation?

d. How many minutes does it take for the secret to be known by:
   1) a group of 32 students
   2) a small town with a population of 16 384 people
   3) a big city with a population of 524 288 people

e. How many people know the secret after:
   1) 20 min?
   2) 25 min?
   3) 33 min?

Worst-kept secret

Time (min)	Calculation	Number of people aware of the secret
0	1×2^0	1
1	$1 \times 2 = 1 \times 2^1$	2
2	$1 \times 2 \times 2 = 1 \times 2^2$	4
3	$1 \times 2 \times 2 \times 2 = 1 \times 2^3$	8
4	▬	▬
5	▬	▬
6	▬	▬
7	▬	▬
8	▬	▬
9	▬	▬
…	…	…
t	▬	

A Petri dish is a very shallow plastic or glass dish with a cover which is used to cultivate cells or micro-organisms.

A technician is testing the effectiveness of an antibiotic ointment by applying a certain amount of antibiotic ointment to a bacterial culture every 24 h. She then observes the number of bacteria through a microscope. The adjacent graph shows the number of bacteria in the culture in relation to the number of applications of the antibiotic ointment.

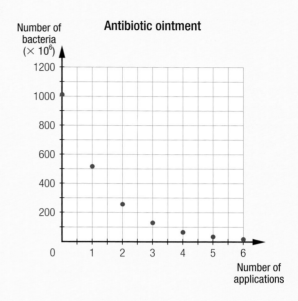

a. In your own words, describe the change in the number of bacteria in relation to the number of applications of the antibiotic ointment.

b. Complete the table below.

Antibiotic ointment

Number of applications	Calculation	Number of bacteria ($\times 10^6$)
0	1024×0.5^0	1024
1	$1024 \times 0.5 = 1024 \times 0.5^1$	512
2	$1024 \times 0.5 \times 0.5 = 1024 \times 0.5^2$	256
3	$1024 \times 0.5 \times 0.5 \times 0.5 = 1024 \times 0.5^3$	
4		
5		
6		
7		
...	...	...
n		

c. What is the initial value in the situation?

d. By what number should you multiply the number of bacteria in order to determine the number of bacteria present after the next application?

e. The graph of the mathematical model related to this situation is a curve. What can be said about the distance between this curve and the x-axis as n increases?

Techno math

A graphing calculator allows you to determine the rule of a function that can be used as a mathematical model in a given situation.

This table of values shows data collected during an experiment relating two variables.

x	y
1	90
2	50
3	22
4	18
5	8
6	6

This screen allows you to enter and edit each of the ordered pairs from the table of values.

Screen 1

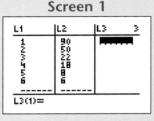

Screen 2

This screen allows you to select the scatter plot as the display mode.

Screen 3

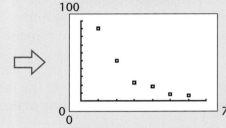

The graphic trend shown by the points corresponds to that of an exponential function.

Screen 4

Using this screen, you can determine the rule of the exponential function whose curve best fits the scatter plot.

Using these two screens, you can obtain the rule of an exponential function and enter the result in the rule editor.

Screens 5 and 6

ExpReg L₁,L₂

```
ExpReg
 y=a*b^x
 a=143.7306783
 b=.5771366719
```

Screen 7

This screen allows you to display the curve of the exponential function on the scatter plot.

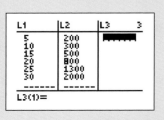

a. In Screen **6**, what do the values a and b represent?

b. Using a graphing calculator and the adjacent data, do the following:

1) Display a scatter plot.

2) Find the rule of an exponential function that can be used as a mathematical model for this situation.

3) Draw the curve on the scatter plot.

knowledge 4·3

EXPONENTIAL FUNCTIONS

An **exponential function** is a function defined by a rule in which the independent variable appears as an exponent.

In the case of an exponential function whose rule is written in the form $f(x) = a(\text{base})^x$ where $a \neq 0$ and where the base is greater than 0 and not equal to 1, the following is true:

- For each unit increase of the independent variable, the dependent variable reflects a sequence in which each term is linked to the next by the same multiplicative factor, the base of the function.

- The graphical representation is a curve that passes through point (0, a) and one of the ends of this curve approaches the x-axis while without ever touching it.

A line that a curve slowly approches but never touches is called an **asymptote**.

Description	Rule	Table of values
E.g. For each unit increase of the independent variable, the corresponding dependent variable is obtained by multiplying the previous value by 3.	$f(x) = 3^x$	(see table)

Table of values:

x	y
-3	$\frac{1}{27}$
-2	$\frac{1}{9}$
-1	$\frac{1}{3}$
0	1
1	3
2	9
3	27

(+1 on x, ×3 on y) — Multiplier that corresponds to the base of the function

Graphical representation

Asymptote

PARAMETER a

For an exponential function whose rule is written in the form $f(x) = a(\text{base})^x$, the variation of **parameter a** generates a change in the vertical scale of the graph.

- The further the value of **a** from 0, the more the curve is vertically stretched.

- The more the value of **a** approaches 0, the more the curve is vertically compressed.

- When the sign of **a** changes, the curve undergoes a reflection over the *x*-axis.

E.g. 1)

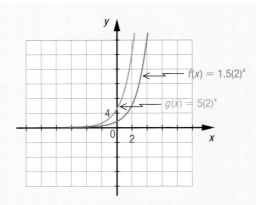

Since 5 is further away from 0 than 1.5, the curve of function *g* is more vertically stretched than that of function *f*.

Also 1.5 is closer to 0 than 5, so the curve of function *f* is more vertically compressed than that of function *g*.

2)

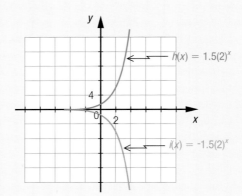

Since -1.5 is opposite to 1.5, the curve of function *h* is a reflection of function *i* over the *x*-axis.

THE BASE OF AN EXPONENTIAL FUNCTION

The value of the **base** of an exponential function affects its graphical representation. For an exponential function whose rule is written in the form $f(x) = a(\text{base})^x$, the following is true:

- When the base is greater then 1, the curve moves away from the *x*-axis as the value of the independent variable increases.

- When the base is between 0 and 1, the curve comes closer to the *x*-axis as the value of the independent variable increases.

 E.g. 1)

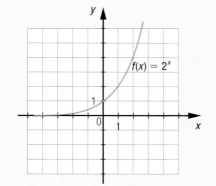

2)

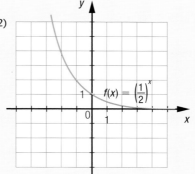

THE RULE OF AN EXPONENTIAL FUNCTION

The rule of an exponential function in the form $f(x) = a(\text{base})^x$ can be determined as follows.

1. Substitute the initial value of the function for parameter **a**.	E.g. On the graph below, the initial value of the function is 3. Thus, the rule is of the form $f(x) = 3(\text{base})^x$. 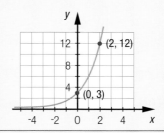
2. Substitute *x* and *y* with the coordinates of a point belonging to the function and that is not located on the *y*-axis.	The curve passes through the point (2, 12). $12 = 3(\text{base})^2$
3. Solve the equation to determine the value of the base of the function.	$12 = 3(\text{base})^2$ $4 = (\text{base})^2$ $2 = \text{base}$
4. Write the rule of the function.	$f(x) = 3(2)^x$
5. Validate the solution.	$12 = 3 \times 2^2$ $12 = 3 \times 4$ $12 = 12$

1 For each of the following, determine the value of x.

a) $x = 3^4$ b) $x = 7^5$ c) $64 = x^3$ d) $625 = x^4$

e) $81 = 9^x$ f) $144 = 12^x$ g) $2^x = 32$ h) $27 = 3^x$

2 Complete the following table.

	Rule of the function	Domain	Range	Base	Initial value	Variation
a)	$f(x) = 3\left(\frac{1}{5}\right)^x$					
b)	$g(x) = 2.5^x$					
c)	$h(x) = 0.7^x$					
d)	$i(x) = 2(\pi)^x$					
e)	$j(x) = -2\left(\frac{2}{3}\right)^x$					
f)	$k(x) = -7(2.6)^x$					

3 For each of the number sequences below, do the following:

1) Complete the sequence.
2) State the rule that allows you to determine the value of a term based on its rank in the sequence.

a) 6, 18, 54, 162, ▨▨▨ , ▨▨▨ , ▨▨▨ , ...

b) 2048, 256, 32, ▨▨▨ , ▨▨▨ , ▨▨▨ , ...

c) 0.0025, 0.025, 0.25, ▨▨▨ , ▨▨▨ , ▨▨▨ , ...

d) 768, 192, 48, ▨▨▨ , ▨▨▨ , ▨▨▨ , ...

4 a) Draw a graphical representation of each pair of functions provided below on the same Cartesian plane.

1) $f(x) = 3^x$ and $g(x) = \left(\frac{1}{3}\right)^x$

2) $h(x) = 3(2^x)$ and $i(x) = -3(2^x)$

3) $j(x) = 3\left(\frac{1}{4}\right)^x$ and $k(x) = -3\left(\frac{1}{4}\right)^x$

b) For each of the graphs in **a)**, indicate the geometric transformation that associates the two curves.

5 Each year, the frog population of a small wooded area declines by 5% in contrast to the previous year. If this wooded area now has 2000 frogs, how many frogs will be present 10 years from now?

Because of the semi-permeability of their skin, frogs are extremely sensitive to pollution. The fluctuations in frog populations provide good indicators of the changes taking place in the environments.

6 Among the tables of values below, which ones represent exponential functions?

A

X	Y₁
1	3
2	12
3	27
4	48
5	75
6	108
7	147

X=1

B

X	Y₂
1	5
2	20
3	45
4	80
5	125
6	180
7	245

X=1

C

X	Y₃
1	6
2	30
3	150
4	750
5	3750
6	18750
7	93750

X=1

D

X	Y₄
-3	0
-2	2
-1	4
0	6
1	8
2	10
3	12

X=-3

E

X	Y₅
-3	-1.375
-2	-2.75
-1	-5.5
0	-11
1	-22
2	-44
3	-88

X=-3

F

X	Y₆
-3	-2.7
-2	-1.2
-1	-.3
0	0
1	-.3
2	-1.2
3	-2.7

X=-3

7 **CREDIT CARDS** Among the options available to finance a purchase, credit cards are the ones that have the highest interest rates. If Diane makes a purchase worth $1,200 with a credit card that has an interest rate of 1.5% each month, how much will she pay in interest if she can only clear her card a year later?

8 Each screen below provides the first three terms in a sequence of numbers obtained using the same process. For each screen, find the value of the 100th term in the sequence.

Screen 1

```
-3.5*2
            -7
Ans*2
           -14
Ans*2
           -28
```

Screen 2

```
7*1.2
           8.4
Ans*1.2
          10.08
Ans*1.2
         12.096
```

Screen 3

```
10*π
      31.41592654
Ans*π
      98.69604401
Ans*π
      310.0627668
```

9 **ARITHMETIC SEQUENCES AND GEOMETRIC SEQUENCES** In an arithmetic sequence, the difference between two consecutive terms is always the same. In a geometric sequence, it is the ratio of two consecutive terms that always remains the same. The following table shows an arithmetic sequence and a geometric sequence.

	+3	+3	+3	+3	+3	+3	+3	+3	+3
Arithmetic sequence	1	4	7	10	13	16	19	22	25	28
Rank	1	2	3	4	5	6	7	8	9	10
Geometric sequence	1	3	9	27	81	243	729	2187	6561	19 683
	×3	×3	×3	×3	×3	×3	×3	×3	×3

a) On the same Cartesian plane, show the relationship between the rank and the value of the terms in each sequence.

b) Identify the type of function associated with each sequence.

c) For each sequence, find the rule of the function that allows you to determine the value of a term according to its rank.

10 Determine the value of the missing terms in the following geometric sequences.

a) ▨, 56, 896

b) 13, ▨, ▨, 1625

c) 4, ▨, 324, ▨, 26 244

d) ▨, 10, ▨, 2560

11 The temperature of the coldest part of a refrigerator varies according to the rule $T = 20(0.75)^x$ where T represents the temperature (in °C) and x represent the time elapsed (in h) since plugging in the refrigerator.

a) Draw a graphical representation of the temperature in relation to time.

b) In the graph obtained in **a)**, do the following:
   1) Determine the equation of the asymptote.
   2) Taking the context into account, explain what the asymptote represents.

c) What is the range of this function?

d) What is the temperature of the refrigerator before it is plugged in?

12 Emily uses the following method to find the rule of an exponential function expressed in the form $y = a(\text{base})^x$ and whose curve passes through points (3, 54) and (5, 486).

1. I'm writing down two equations using given points. $486 = a(\text{base})^5$ and $54 = a(\text{base})^3$ 2. Now, I'm dividing the two equations and simplifying them. $\dfrac{486 = a(\text{base})^5}{54 = a(\text{base})^3}$ $9 = (\text{base})^2$ 3. And now, I'm finding the base using a square root. $3 = \text{base}$	4. By replacing the base and the coordinates of a point expressed in the following form, $y = a(\text{base})^x$, I can find the value of a. $486 = a(3)^5$ $486 = 243a$ $2 = a$ 5. The rule for the exponential function expressed in the form $y = (\text{base})x$ and whose curve passes through points (3, 54) and (5, 486) is $y = 2(3)^x$.

Using the same method, find the rule for the exponential functions expressed in the form $y = a(\text{base})^x$ and whose curve passes through points:

a) (1, 24) and (4, 5184)

b) (3, 10, 125) and (-1, 2)

c) (4, -81) and (7, -2187)

d) (-3, 16) and (2, 0.5)

13 **CANADIAN POPULATION** In Canada, a census is officially taken every five years. When the 2006 census was taken, the population of Canada was 31 241 030; this represented a 5.4% increase compared to the 2001 census.

a) What was the population of Canada in 2001?

b) By how many people did the population of Canada increase between 2001 and 2006?

c) If the population of Canada increases by 5.4% whenever a census is taken, what should be the population of Canada in 2026?

Since the data are highly detailed, and the questions asked remain the same from one census to another, the information on the Canadian population is very precise.

d) Explain why the value found in **c)** is an estimate.

14 **NEPER'S CONSTANT** The letter "e" is used to refer to Neper's constant, which has an approximate value of 2.718 281 828 459. This constant, like the constant π, is an irrational number found in a number of mathematical formulas. The number e can be expressed with the sequence:

$$e = 2 + \frac{1}{2} + \frac{1}{3 \times 2} + \frac{1}{4 \times 3 \times 2} + \frac{1}{5 \times 4 \times 3 \times 2} + \frac{1}{6 \times 5 \times 4 \times 3 \times 2} + \cdots$$

In finance, the constant e is used to calculate compound interest of an investment using the rule $A = Pe^{rt}$ where A represents the final value of the investment, P is the initial amount and r is the annual interest rate for a period of t years.

a) Use a graphical representation to represent a rule of the form $A = Pe^{rt}$ for an investment of $20,000 at an interest of 4.5%.

b) Based on this model, what is the final value of an investment of $20,000 at an interest rate of 4.5% for a period of 10 years?

15 Consider the function $f(x) = 5(2)^x$.

a) Complete the adjacent table of values.

b) Draw a graphical representation of function f.

c) Specify its domain and range.

d) Draw a graphical representation of the inverse of function f.

e) Provide the domain and the range of the inverse of function f.

f) Is the inverse of function f a function? Explain your answer.

x	f(x)
-3	
-2	
-1	
0	
1	
2	
3	

16 A person invests $5,400 in a guaranteed investment certificate with an annual interest rate of 3.6%. What will be the value of the investment after 10 years if no withdrawals are made?

17 The percentage P of light passing through a number n of sheets of transparent acrylic plastic can be calculated using the rule $P = 100(0.985)^n$.

Acrylic plastic is used for its properties of transparency and strength that are superior to those of glass. It is marketed under the name *Plexiglas*.

a) What percentage of light is blocked:
   1) by one sheet of acrylic plastic?
   2) by 10 sheets of acrylic plastic?

b) What percentage of light passes through:
   1) one sheet of acrylic plastic?
   2) 10 sheets of acrylic plastic?

18 **WORLD POPULATION** In a report published in 2004, the United Nations calculated the population growth rate to be 1.22% for each year. One year later, the same organization estimated the world population at 6 454 000 000.

a) According to this information, what was the world population in 2004?

b) If the growth rate remains the same, what rule allows you to calculate the world population in relation to the number of years elapsed since 2004?

c) What will the world population be in 2050?

19 The adjacent table of values provides information about a bacterial culture.

a) Determine the rule that would allow you to calculate the number of bacteria as a function of time.

b) When will the number of bacteria reach:
   1) 192 000?   2) 3 072 000?

Bacterial culture

Time (h)	Number of bacteria
0	3000
1	6000
2	12 000
3	24 000
4	48 000

20 A couple wants to invest $25,000.

> These are two medium-term investment plans, Plan **A** with an annual interest rate of 6% compounded annually, and Plan **B** with an annual interest rate of 6% compounded every six months.

a) According to Plan **A**, what will the value of this investment be at the end of one year if no withdrawals are made?

b) What percentage of the initial investment does the amount calculated in **a)** amount to?

c) Based on Plan **B**, what will the value of the investment be at the end of one year if no withdrawals are made?

d) What percentage of the initial investment does the amount calculated in **c)** amount to?

e) Complete the following table.

Plan A				Plan B			
Time (month)	Time (years)	Calculation	Value of investment ($)	Time (month)	Time (years)	Calculation	Value of investment ($)
0	0	$25\,000(1.06)^0$	25,000	0	0	$25\,000(1.03)^0$	25,000
				6	0.5	$25\,000(1.03)^1$	
12	1			12	1	$25\,000(1.03)^2$	
				18	1.5		
24	2			24	2		
				30	2.5		
36	3			36	3		
				42	3.5		
48	4			48	4		
...	...	...	...	...	...	...	...
	x				x		

f) Which of the two investments is more profitable? Explain your answer.

This section is related to LES 9.

PROBLEM   Horizontal Falls

The *Horizontal Falls*, located in Talbot Bay, are one of the natural wonders of Australia. Though they are called "falls," they are, in fact, an intense marine current that rises between two gorges too narrow to allow an easy flow of water. As the tide rises or falls, a difference in water level occurs on either side of these gorges; this generates a phenomenon resembling waterfalls that can reach up to 4 m in height. At high tide, the falls flow towards the coast while the opposite occurs at low tide. The phenomenon is most impressive when the tide is at its highest, and this is the best time to observe these falls.

The table and the graph below are used to predict the height of the tide based on the time of day.

January 12 tides	
Time	Height (m)
6:15 a.m.	10.85
7:10 a.m.	10.1
8:05 a.m.	8.6
9:00 a.m.	6.35
9:55 a.m.	4.1
10:50 a.m.	2.6
11:45 a.m.	1.85
12:40 p.m.	2.6
1:35 p.m.	4.1
2:30 p.m.	6.35
3:25 p.m.	8.6
4:20 p.m.	10.1
5:15 p.m.	10.85

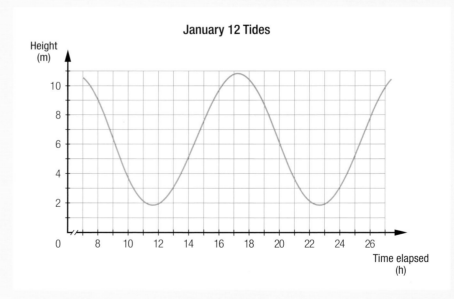

January 12 Tides

At what time in the afternoon on January 19 would you recommend going to see these falls?

ACTIVITY 1 What's to come?

Weather forecasts are possible through interpretation of satellite images, radar images and data collected from a vast network of weather stations. The data collected by these stations include temperature, relative humidity, atmospheric pressure, wind speed and direction.

The study of variations in atmospheric pressure allows meteorologists to track weather systems such as atmospheric lows that cause bad weather.

The graph below shows the atmospheric pressure recorded at the same location during the course of one day.

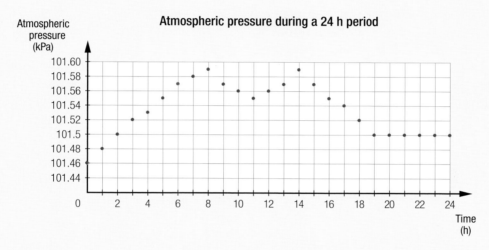

a. The scatter plot representing this situation is made up of distinct parts. What type of function can you associate with each of the following intervals?

1) [0, 8] h   2) [8, 14] h

3) [14, 19] h   4) [19, 24] h

b. What type of function can you associate with this situation?

c. Draw the curve of best fit for each part of this function.

d. According to the model obtained in c., determine the atmospheric pressure at:

1) 2:30 a.m.   2) 8:25 a.m.

3) 2:45 p.m.   4) 6:15 p.m.

Stevenson shelters are designed to protect the data sensors from both direct and indirect light. The white colour allows for reflection of the sun's rays and the shutters promote air circulation.

Income tax is a mandatory deduction on a person's income, paid annually to cover State or community expenses.

The taxation system used in Canada is called graduated income tax. However, certain countries, like Iceland, have chosen to adopt the proportional income tax system, in other words, a single taxation rate for everyone regardless of income.

In Canada, federal taxation rates generally vary according to brackets based on an individual's taxable income. The information below concerns federal private income tax rates for the year 2008:

Federal income tax

Taxable income	Taxation rate (%)
Over $0 without exceeding $37,885	15
Over $37,885 without exceeding $75,769	22
Over $75,769 without exceeding $123,184	26
Over $123,184	29

For example, an individual whose taxable income is $65,000 will pay 15% federal income tax on the first $37,885 of income and 22% on the additional $27,155.

a.
1) Construct a graph that represents the data from the table.
2) What type of function can you associate with this situation?
3) Determine the domain and range of this function.

b. Calculate the amount of federal income tax paid by an individual whose taxable income is:
1) $35,000
2) $75,769
3) $145,750

c. What would be the taxable income for an individual:
1) whose income tax rate on the last portion of their taxable income is 26%?
2) who paid federal income tax in the amount of $9,300?

ACTIVITY 3 — Wind turbines

Since ancient times, windmills have been used to convert wind energy for milling grain or pumping water. The first aerogenerators, commonly called "wind turbines," appeared at the end of the 19th century. These devices make it possible to convert wind energy into electricity.

Wind turbine

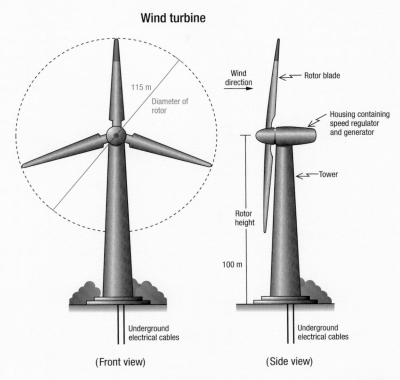

(Front view)   (Side view)

Thea watches the red tip of one of the rotor blades of a wind turbine. The rotor blades of this wind turbine revolve at a speed of 20 revolutions per minute. At the start of her observations, this tip is at its lowest position.

a. Construct a graph representing the height (in m) of the red tip above ground level in relation to time (in s) during the first 10 seconds.

b. What type of function can you associate with this situation?

c. Determine the domain and range of this function.

d. During the first 20 seconds, at what times is the red tip at its initial height?

e. Which of the function's properties will be altered if:
1) the diameter of the rotor is reduced to 75 m?
2) the rotor blades revolve at a speed of 15 revolutions per minute?

Wind energy is a renewable type of energy; it is naturally regenerated and is inexhaustible. Other examples of renewable energy include tidal power, hydropower and geothermal energy.

Techno math

A graphing calculator allows you to graphically represent piecewise functions.

These two screens allow you to select inequality symbols and logical connectors among other options.

Screen 1

Screen 2

Using inequality symbols and logical connectors, it is possible to limit and define the domain of a function.

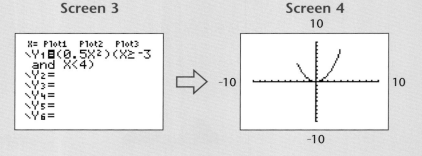

Screen 3

Screen 4

The adjacent screen displays an example of how to enter the rule of a piecewise function.

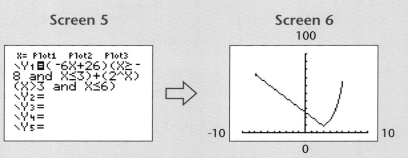

Screen 5

Screen 6

a. Based on Screens **3** and **4**, what is:
   1) the domain of the function?
   2) the range of the function?

b. Based on Screens **5** and **6**, what is:
   1) the domain of the first part of the function?
   2) the domain of the second part of the function?
   3) the domain of the piecewise function?
   4) the range of the piecewise function?

c. Using a graphing calculator, do the following:
   1) Display the function whose rule is $y = 2(3)^x$ and domain is $]1, 4]$.
   2) Display a piecewise function where the first piece is defined by the rule $y = 0.6x^2$ over the domain $[-8, 4]$ and the second piece is defined by the rule $y = x - 30$ over the domain $]4, 9[$.

knowledge 4.4

STEP FUNCTION

A step function is a function that is constant for certain intervals and then varies suddenly at certain values of the independent variable, called "critical values." The graphical representation of this function is made up of horizontal segments. At the extremities of each segment, a solid or clear dot is used to designate a pair of values that may or may not belong to the function.

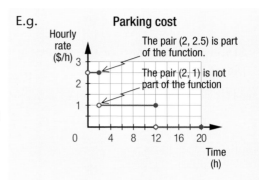

E.g. **Parking cost**

The pair (2, 2.5) is part of the function.

The pair (2, 1) is not part of the function

The cost to park a car in a parking lot varies in relation to the duration. The hourly rate is $2.50 for the first two hours. Then, it decreases to $1/h for the 10 subsequent hours after which there are no additional charges. The parking lot is open 20 h/day.

PERIODIC FUNCTION

A function is said to be periodic when its graphical representation consists of a regularly recurring "pattern." The interval between the x-values located at the extremities of this recurring "pattern" is the period of this function.

E.g. **Oscillation of a mass**

The movement of a mass suspended from a spring oscillating vertically without friction can be modelled by a periodic function.

It is possible to conclude that the mass returns to its original position every 2 s. The period of this function therefore is 2 s.

PIECEWISE FUNCTION

A piecewise function is a function made up of several functions defined by different intervals within the domain. The parts that make up such a function may come from one or several families of functions.

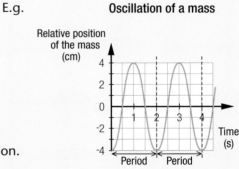

E.g. **Speed of a car**

The speed of a car that is accelerating at a constant rate, maintaining its speed and later slowing down can be modelled by a piecewise function.

This function is made up of a second-degree polynominal function, zero-degree polynominal function and a first-degree polynominal function.

practice _{4.4}

1 State whether each of the following graphs represents a step, periodic or piecewise function.

a)

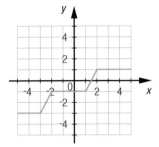

b)

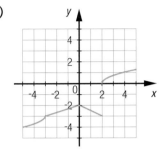

c)

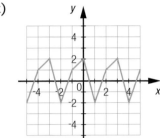

d)

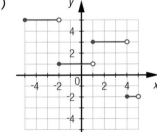

2 State whether each of the following situations represents a periodic, piecewise or step function.

a) A commercial enterprise offers its customers a free ticket to a show in return for each $150 spent on purchases. Consider the relation between the total amount of a customer's purchases and the number of show tickets received.

b) An all-terrain vehicle travels a few kilometres at a constant speed, slows down, then stops for a few minutes to fill up with gas. Consider the relation between time elapsed and the distance covered by this vehicle.

c) A metal ball is suspended on the end of a spring that is oscillating with a regular motion above the ground. Consider the relation between time elapsed since the start of the motion and the height of the ball in relation to the ground.

All terrain vehicules (ATVs) have four-wheel drive allowing them to drive over all types of terrain. These vehicules are very popular with the military, foresters and farmers.

3 For each of the functions below, determine:

1) the type of function represented
2) the initial value of the curve
3) the value of y when $x = 10$
4) the value(s) of x when $y = 4$

a)
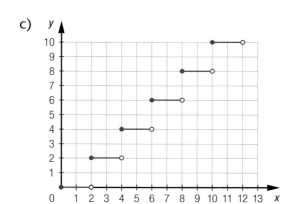

b)
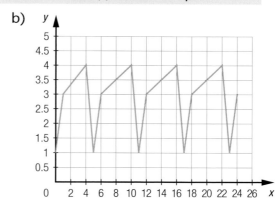

c)
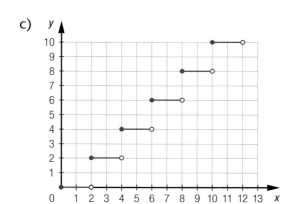

d)
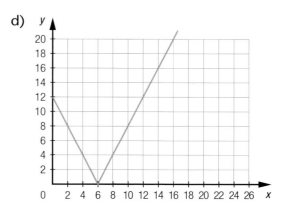

4 A parking lot advertises the adjacent rates.

a) Construct a graph representing this situation if a car is parked on this lot for any period of time from 0 to 12 h.

b) How much does it cost to park a car in this lot for a period of 7 h?

c) Is it possible that a person might pay $10 to park a car in this lot? Explain your answer.

First hour:
$5

Each subsequent hour:
$2

Maximum daily rate:
$15

5 Consider the adjacent graphical representation of a periodic function.

a) What is the period of this function?

b) Determine the range of this function.

c) Calculate:

1) $f(-3)$ 2) $f(12)$ 3) $f(-20)$

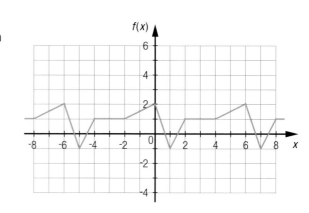

6 A computer technician is testing the performance of a computer's file retrieval program. The adjacent graph shows the results of the analysis.

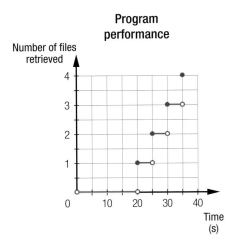

Program performance

a) In this situation, what do the critical values represent?

b) How many files did this program retrieve during this search?

c) Determine the domain and range of this function.

d) Determine the zeros of this function and interpret them in relation to the context.

7 Consider the following description of a stock's performance over the past 12 months: At the start, the stock was valued at $1.00 for each share. Over the first 4 months, the share value increased regularly at the rate of $0.50/month; after that it stabilized for 2 months. At the beginning of the 7th month, share values spiked by $1.00 and stabilized for the rest of the month. The same phenomenon, followed by a one-month stabilization period, occurred as soon as the 7th month was over. At the end of the 8th month, the share value spiked again by $1.00, then began to decrease regularly at the rate of $0.50/month until the end of the 12-month period.

a) Construct a graph representing the stock's performance.

b) What type of function can you associate with this performance?

c) What was the share value at the end of 12 months?

d) During which month(s) was the share valued at $5.00?

8 Consider three experiments with a marble:

Experiment ①	**Experiment ②**	**Experiment ③**
A marble rolls without friction in a semi-spherical bowl 15 cm in diameter. The marble is released at the edge of the bowl and always passes through its centre.	A marble rolls on an inclined board 30 cm in length; it then continues its path on the floor. At the start, the marble is placed at a height of 15 cm.	A marble rolls down a staircase comprised of 3 steps. The depth and height of each step measure 30 cm and 25 cm, respectively. The marble is released from the top of the first riser.

Consider the height of the marble (in cm) in relation to time (in s) and do the following:

a) Determine the function that best models each of these situations.

b) Draw the sketch of the graph corresponding to each of these situations.

9 Is there a periodic function whose inverse is also a function? Explain your answer.

10 Rental rates for a fully-equipped river kayak are $25/day for a 1 to 6-day rental and $18.75/day for 7 days or more. The kayak can only be rented for full days.

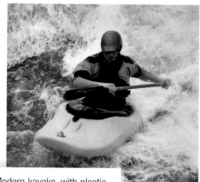

a) Construct a graph representing the rental rates for 0 to 15 days.

b) If an individual wants to go kayaking for 5.5 days, is it cheaper to rent the kayak for 6 days or for 7 days? Explain your answer.

c) If an individual pays $150 to rent a kayak, for how many days can he use the kayak?

Modern kayaks, with plastic or fibreglass hulls, are based on traditional Inuit craft that had hulls made of seal skin.

11 **IRONMAN TRIATHLON CANADA** On August 26, 2007, Jonathan Caron finished second in the Ironman Triathlon Canada, in Penticton, British Columbia. The table below provides information on his triathlon.

Triathlon

Event	Distance (km)	Mean speed (m/s)
Swimming	3.8	1.23
Cycling	180	10.37
Running	42.2	3.99

a) Draw a graphical representation showing:
  1) mean speed in relation to time
  2) distance in relation to time

b) What discipline did Jonathan Caron practice 63 min after the start?

c) How much time did it take Jonathan Caron to complete this triathlon?

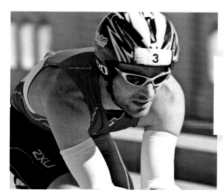

Ironman competitions are the longest triathlon races. Each year, in October, participants who have qualified in Ironman races around the world are eligible for the Ironman World Championship held in the United States.

12 LOW CAPACITY FLUSH SYSTEMS   Before 1980, most toilets had a flush system requiring 20 L of water. Today, most flush systems use 13 L of water while some very low capacity systems use only 6 L. The diagram below shows the amount of water in a toilet's tank after flushing.

a) At what rate does the tank:

   1) empty?

   2) refill?

b) What is the amount of water:

   1) 3 s after flushing?

   2) 15 s after flushing?

c) How many seconds after flushing does the tank contain 8 L of water?

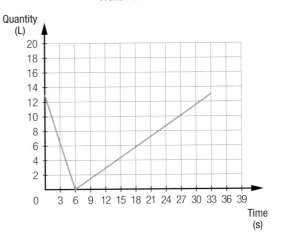

Water in the tank

13 An individual microwaves popcorn for 2 min. The adjacent table shows the speed at which the corn kernels pop in relation to cooking time.

Popcorn, found in practically all movie theatres, is associated with going to the movies. It is also used to wrap fragile items, replacing polystyrene which is not a biodegradable packing product.

Popping the corn

Cooking time (s)	Mean number of kernels of corn popped
[0 - 10[	3
[10 - 20[	40
[20 - 30[	185
[30 - 40[	390
[40 - 50[	505
[50 - 60[	500
[60 - 70[	530
[70 - 80[	515
[80 - 90[	430
[90 - 100[	350
[100 - 110[	300
[110 - 120[	30
[120 - 130[	0

a) Represent this situation using a graphical representation.

b) How many kernels of corn have popped:

   1) in 45 s?

   2) in 72 s?

   3) when the cooking time has finished?

c) When will 10 000 kernels have popped?

14 **SLEEP ENCEPHALOGRAM** A sleep encephalogram is a recording of an individual's physiological functions and is used to study his or her state of sleep. The sleep encephalogram below was recorded during a normal adult's 8 h of sleep:

Sleep encephalogram

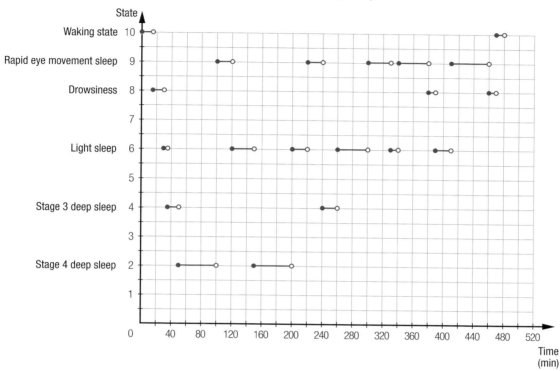

Considering that dreams occur mostly during the period of rapid eye movement (REM) sleep, for how many hours did this person dream?

15 A retail salesman receives a base salary of $340/week. In addition to this salary, he receives a commission calculated on the basis of the weekly amount of his sales. The adjacent table gives a summary of this situation.

Determining commissions

Amount of sales ($)	Commission ($)
1,500 or less	0
Over 1,500 but not exceeding 2,000	20
Over 2,000 but not exceeding 2,500	40
Over 2,500 but not exceeding 2,750	60
Over 2,750 but not exceeding 3,000	80
3,000 or more	100

a) Construct a graph representing the total salary in relation to the amount of sales.

b) What is the total salary of a salesperson whose sales amount to:
   1) $1,200?
   2) $2,225?
   3) $3,000?
   4) $3,450?

c) What is the minimum amount of sales required to earn at least $400?

d) Can a salesperson earn exactly $350? Explain your answer.

Chronicle of the
past
The history of electricity

André-Marie Ampère

André-Marie Ampère (1775-1836)

André-Marie Ampère was a French mathematician and physicist born in Lyon on January 20, 1775. He never went to school because his father insisted on teaching André-Marie himself. At the early age of 13, André-Marie took an interest in mathematics which quickly became a genuine passion for him. In 1801 he became a professor of mathematics and physics in Bourg and Lyon. After his death in 1836, his name was given to the unit of measure for the intensity of an electrical current: the ampere.

Ampère's work

Ampère worked on many projects during his lifetime. His research mainly concerned the fields of probability, magnetism and electricity. He was the first to form a theory of electromagnetism. Based on this theory, he and François Arago later invented the electromagnet. He also proposed the existence of electric currents. He is, in fact, the first to have used the term "electric pressure." At the time, Ampère was known as the "Newton of electricity."

Georg Ohm (1789-1854)

Ampère's discoveries would lead another scientist, Georg Ohm, to formulate a law connecting the intensity and the voltage of an electric current. Ohm's law verifies that, in an electric circuit, the voltage V in (volts) equals the product of the resistance R (in ohms) by the intensity I of the current (in amperes). The scatter plot below represents the data collected by Ohm in the course of an experiment on electricity.

The principle of the electromagnet is the following: a coil of conductive wire is attached to a source of current. The current circulating in this coil generates a magnetic field similar to that of a magnet.

Resistance of 2.5 ohms

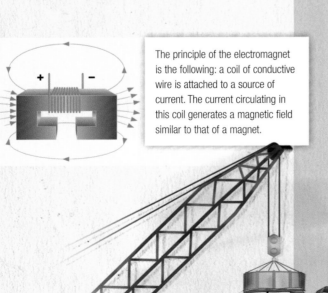

An electromagnet can be strong enough to lift several tons.

James Prescott Joule (1818-1889)

In 1841, James Prescott Joule, a British physicist, defined a law verifying that the power P dissipated (in watts) in an electric circuit is equal to the product of the resistance R (in ohms) by the intensity I of the current (in amperes) squared. The power dissipated in an electric circuit translates into heat and is called the Joule effect. Below is the laboratory data collected while studying the Joule effect on an aluminum wire with a resistance of 10 000 ohms:

Joule mainly worked to establish the link between mechanical energy and thermal energy. The unit of work, heat and energy in the international system of units, the joule (J), bears his name.

Joule effect

Intensity (A)	0	0.5	1	1.5	2	2.5	3	3.5	4	4.5	5
Power dissipated (kW)	0	2.5	10	22.5	40	62.5	90	122.5	160	202.5	250

High voltage power lines are used to transport electricity as a means of diminishing energy loss as described by the Joule effect.

1. Based on the scatter plot representing the data collected by Ohm, answer the following:

a) What type of function can you associate with this situation?

b) What is the intensity of the current when the voltage is 20 V?

c) What change would occur to the scatter plot if the resistance was increased to 5 ohms?

2. a) Construct a graph representing laboratory results during the study of the Joule effect on an aluminum wire.

b) What type of function can you associate with this situation?

c) What change will occur to the curve associated with this situation if the value of the wire's resistance is decreased?

In the workplace

Deep-sea divers

Training

Professional deep-sea divers are people who have a college certificate in professional deep-sea diving. As a prerequisite to this training, candidates must earn a vocational or secondary school diploma, a Class A recreational diving certificate and pass a medical examination.

A demanding trade

Professional deep-sea divers are expected to perform a great variety of underwater tasks such as welding, cutting, repairing, inspecting, salvaging wrecks and performing rescue operations. They may also teach deep-sea diving.

In 2006, of the 135 professional deep-sea divers in Québec, there were 6 women practising this trade.

The main concern: safety

The work of professional deep-sea divers is very demanding, both physically and mentally, and can sometimes be very dangerous. A high level of physical fitness, the ability to remain calm and complete commitment to safety are basic requirements for the practise of this trade.

Hazards associated with sudden pressure variations

At sea level, the human body is subjected to a normal atmospheric pressure of 1 bar. However, for divers, the mass of water is added to this pressure at the rate of around 1 bar for every 10 m of depth.

As divers descend, they must use various balancing techniques in order to avoid dizziness or permanent damage to their eardrums.

Underwater, the diver's blood fills with tiny bubbles of nitrogen, a gas contained in the compressed air he or she is breathing. As the diver ascends these bubbles of nitrogen become larger, and if the diver does not take the time to eliminate these bubbles during the ascent, he or she faces potential neurological injuries, balance problems or cardiac disorders. An overly speedy ascent can sometimes result in death.

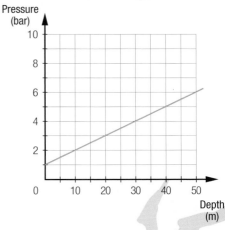

Change in pressure while executing a dive

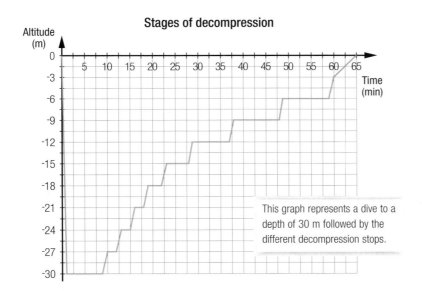

Stages of decompression

This graph represents a dive to a depth of 30 m followed by the different decompression stops.

Sometimes the decompression stops must continue above the surface. In such cases, divers must go into a hyperbaric chamber.

1. What is the pressure sustained by a diver at a depth of:

a) 35 m?

b) 60 m?

2. If a diver must comply with a 5-min decompression stop every 3 m during ascent, how much time will it take for the ascent from a depth of 45 m if the time of ascent between these steps is 1 min?

3. A diver has gone through 14 decompression stops in compliance with one step every 3 m. What was the pressure sustained by this diver when she was at the bottom of her dive?

overview

1 What type of function can you associate with each of the following situations?

a) A nurse checks the curve representing a patient's heart rate on a cardiac monitor.

b) The quantity of blue algae in a lake doubles each month. Consider the relation between time and the quantity of blue algae in the lake.

c) A basketball player makes a free throw. Consider the relation between time and the height of the ball.

d) A salesman's salary amounts to 12% of the total amount of his sales. Consider the relation between the total amount of sales and the salesman's salary.

Blue algae have lived in aquatic ecosystems for billions of years. They become a problem when they proliferate excessively because some species can pose health hazards for animals and humans.

2 What type of function can you associate with the following curves?

a)
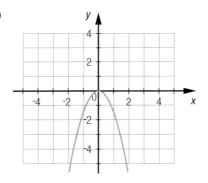

b)
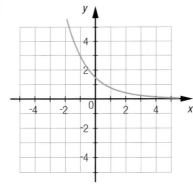

c)
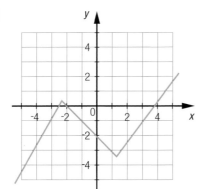

d)
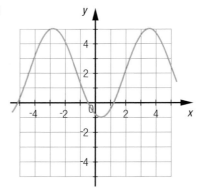

e)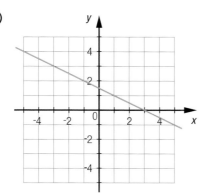

f)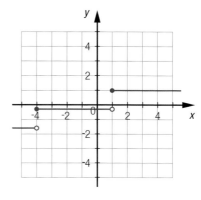

3 For each of the graphs below, do the following:

a) Determine:

    1) the domain and the range   2) the zero(s)

    3) the maximum and the minimum   4) the initial value

b) Determine the type of function.

c) Indicate whether the inverse is a function. Explain your answer.

Graph ①

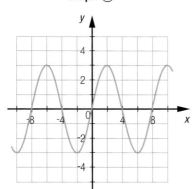

Graph ②

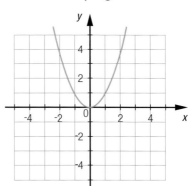

Graph ③

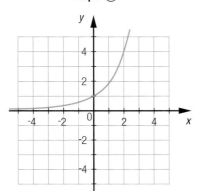

Graph ④

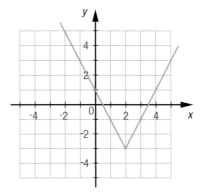

4 Consider the following numerical sequence: 1, 2, 4, 8, 16, 32, 64, 128

Consider the relation between each term's rank and its value in the sequence. What type of function is associated with this situation? Explain your answer.

5 Find the rule of each of the following functions.

a)

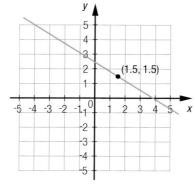

b)

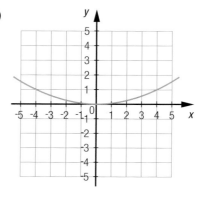

c)

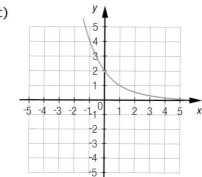

d)

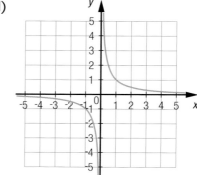

6 For each of the tables of values below, do the following:

a) Construct a graph representing the ordered pairs in a scatter plot.

b) Determine the type of function that best represents the scatter plot.

1)

x	-5	-4	-3	-2	-1	0	1	2	3	4	5
y	-4	-4	-2	-2	0	0	2	2	4	4	6

2)

x	-5	-4	-3	-2	-1	0	1	2	3	4	5
y	3	5	7	9	9	9	10	14	23	39	64

3)

x	-5	-4	-3	-2	-1	0	1	2	3	4	5
y	10	8	6	4	2	0	-2	0	2	4	6

When it was first founded, a children's vacation camp had 3 groups of 5 children and one leader for each group. Every 3 years, 5 more children joined the camp and formed a new group for which a new leader was hired.

a) Complete the following table.

Children's vacation camp

Time since the camp was founded (years)	Number of children	Number of groups	Number of leaders	Number of children for each leader
0	15	3	3	5
1				
2				
3				
4				
5				
6				
7				
8				
9				
10				

b) Draw a graph representing:
   1) the number of children as a function of time
   2) the number of groups as a function of time
   3) the number of children for each leader as a function of time

c) Twenty years after the camp was first founded, how many:
   1) children attend this camp?
   2) groups attend this camp?

d) What type of function can be associated with the change in the number of children for each leader?

Each summer, Camp Papillon, located in the Laurentians, welcomes handicapped children and provides them with outdoor recreational activities. Regardless of their handicap, these campers participate in a variety of activities including fishing, camping, swimming, archery, astronomy and hiking.

8 For each graph below, determine the rule of the function associated with it.

a)

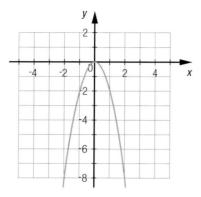

b)

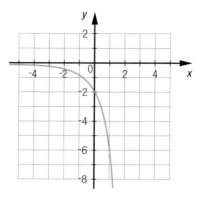

9 The following diagram represents a periodic function:

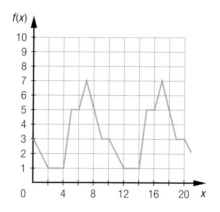

a) What is the period of this function?

b) Determine:

   1) $f(13)$ 2) $f(24)$ 3) $f(41)$

10 The table below represents a summary of the total cost displayed on a taximeter in relation to the distance travelled.

Taximeter

Distance (km)	0	0.25	0.5	0.75	1	1.25	1.5	1.75	2	2.25	2.5	2.75	3
Cost ($)	3.15	3.15	3.15	3.15	4.45	4.45	4.45	4.45	5.75	5.75	5.75	5.75	7.05

a) Represent this situation using a graph.

b) What is the total cost of a taxi ride of 9.7 km?

c) If the total cost of a taxi ride is $10.95, how many kilometres were travelled?

11 The managers of a festival want to give away an amount of money in a draw. This amount will be distributed equally among the winners. The adjacent table of values provides information about this draw.

a) What is the total amount of prize money that will be given away during this draw?

b) Draw a scatter plot representing this situation.

c) What type of function can you associate with this situation?

d) If there are 15 winners, what will each person's share be?

Draw

Number of winners	Each winner's share ($)
1	3600
2	1800
3	1200
5	720
6	600
8	450
12	300

12 The adjacent diagram shows the different stages in a training session for a running competition.

a) How long should a runner, in this training session, run at a rate of:

   1) 4.5 min/km?

   2) 5.5 min/km?

b) How long is this training session?

c) What is the total distance covered by a runner in this training session?

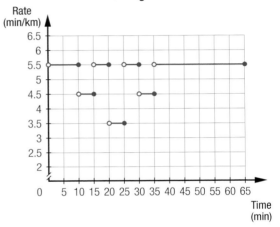

Training session

13 Each year, the number of water-lilies in a pond are counted. The table below provides the results of this report.

Water-lilies in Nuthatch Pond

Time (years)	Number of water-lilies
0	25
1	48
2	112
3	200
4	398
5	804

In the last years of his life, Impressionist painter Claude Monet (1840-1926) painted several hundred paintings of water-lilies, all entitled "Nympheas."

a) Construct a graph representing this situation.

b) What type of function can you associate with this situation?

c) When will there be 3500 water-lilies in this pond?

14 Below is the graphical representation of a step function:

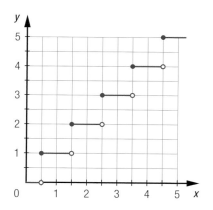

a) Complete the following table of values.

x	0	0.1	0.25	0.4	0.49	0.5	0.6	0.75	0.9	0.99	1	1.1	1.25	1.4	1.49	1.5	1.6	1.75	2
y																			

b) In your own words, describe the operation performed by the function represented above.

15 **ADRENALINE** When people experience danger or a stressful situation, a hormone, adrenaline, is secreted in the body by the adrenal glands. When this substance circulates in the body, heart rate and blood pressure increase. The table below provides information on the quantity of adrenaline produced when a person tries out a new ride in an amusement park.

Adrenaline production

Elapsed time from the start of the ride (s)	5	10	15	20	25	30	35	40	45	50	55	60	65	70	75
Concentration of adrenaline (mg/mL)	10	13	15	19	22	24	25	24	23	24	20	16	11	8	5

What would be the amount of adrenaline in the person's body at start of the ride?

The very first roller coasters were built in Russia in the 17th century. These were huge wooden slides covered with ice that people madly slid down on toboggans. The concept was very popular and was rapidly exported and improved. Wheels were added to the toboggan and later rails to improve safety. Although they are terrifying, modern roller coasters are very safe.

16 An individual who builds scale models of sailboats uses the diagram below as a pattern for the mainsails of the models.

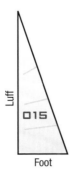

The mainsail is the principal sail on a sailboat's mainmast.

On the pattern, the length of the luff is three times that of the foot.

a) Complete the following table.

Mainsail

Length of the foot (cm)	Length of the luff (cm)	Area of the sail (cm²)
2		
4		
6		
8		
10		

b) Draw a graphical representation of:
   1) the length of the luff in relation to the length of the foot
   2) the area of the sail in relation to the length of the foot

c) Determine the rule that allows you to calculate:
   1) the length of the luff in relation to the length of the foot
   2) the area of the sail in relation to the length of the foot

d) What must be the width of the foot if the area of the sail is 337.5 cm²?

17 The adjacent graph represents the height of a golf ball in relation to time.

a) What function can you associate with this situation?

b) What was the maximum height reached by the ball?

c) At what moment will the ball fall back to the ground?

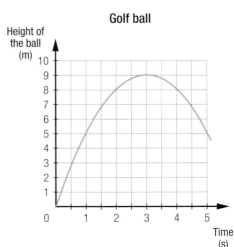

bank of problems

18 **CANADIAN POPULATION** The adjacent table indicates the population of Canada according to census data from 1861 to 2001.

According to this data, estimate what Canada's population will be in 2026.

19 **100-M RACE** The 100-m race is an athletic event consisting of a sprint across a straight track. An athlete who holds the 100-m world record is called "the fastest man on earth" or "the fastest woman on earth." The table below shows the progression in 100-m world records for women.

Population growth

Year	Population (in thousands)
1861	3230
1871	3689
1881	4325
1891	4833
1901	5371
1911	7207
1921	8788
1931	10 377
1941	11 507
1951	13 648
1961	18 238
1971	21 568
1981	24 820
1991	28 031
2001	31 021

Women's 100-m records

Year	1922	1923	1925	1928	1932	1935	1936	1952	1955	1965	1968	1973	1977	1983	1984	1988
Time (s)	12.8	12.7	12.4	12	11.9	11.8	11.5	11.4	11.3	11.2	11	10.9	10.88	10.79	10.76	10.49

Two people propose different models to represent the progression of the world records. Below are the two graphical representations:

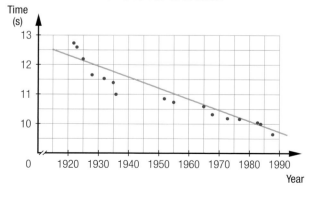

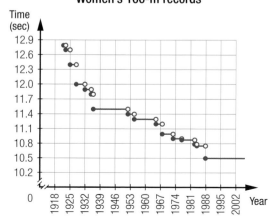

Which of these two models is most appropriate for this context? Explain your answer.

20 While treating a patient, a doctor looks for a medication that will increase the amount of adrenaline in the patient's bloodstream. He also wants the effects of the medication to last at least 90 min and at most 150 min after being administered. The graphical representation below displays the curve associated with the increase of the quantity of adrenaline in relation to time caused by a medication.

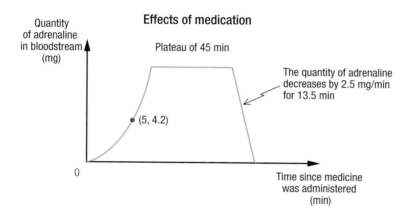

Effects of medication

Quantity of adrenaline in bloodstream (mg)

Plateau of 45 min

The quantity of adrenaline decreases by 2.5 mg/min for 13.5 min

(5, 4.2)

0

Time since medicine was administered (min)

An allergy is a sensitivity to a normally inoffensive substance. Some allergy sufferers can have a serious reaction called *anaphylactic shock* when they come into contact with certain allergens. To avoid such a severe reaction, allergy sufferers must carry an epi pen with them at all times. This way they can prevent a severe reaction by immediately injecting themselves with epinephrin.

Determine if the medication meets the doctor's requests.

21 MALTHUSIAN CATASTROPHE When a population's expansion goes unchecked to the point where numbers exceed natural resources, a collapse of the population occurs. This phenomenon, called a Malthusian catastrophe, applies not only to population growth, but also to the economy. The adjacent table shows the food resources of a small island on which a population of 9 reindeer was introduced.

If the reindeer population increases by 5% each year, how much time will it take before you can expect a Malthusian catastrophe to occur?

Number of reindeer that can subsist on the island's resources

Time (years)	Number of reindeer
0	16
10	20
20	29
30	31
40	36
50	46

British economist, Thomas Robert Malthus (1766-1834), became famous when he published his work entitled *An Essay on the Principle of Population* in 1798. His "Malthusian Doctrine" is based on the principle that if population growth is faster then the production of substances needed for survival, an imbalance occurs, which could result in famine.

VISI⑤N

From right triangles to trigonometric relations

How can you calculate the area of an irregular shaped lot? How was the distance from the Earth to the Moon calculated or the height of the peaks of the Himalaya measured? In what way can you determine your position without using a GPS? In "Vision 5," you will discover the relationships that exist between the measures of the angles and the lengths of the sides of a right triangle. You will find relationships in all types of triangles. You will also learn to use this new knowledge in various practical and concrete applications such as calculating areas or finding missing measurements.

Arithmetic and algebra	Geometry	Statistics	Probability
	• Trigonometric ratios in a right triangle: sine, cosine and tangent • Finding missing measurements • Relations in any triangle: the sine law, Hero's formula and the trigonometric formula		

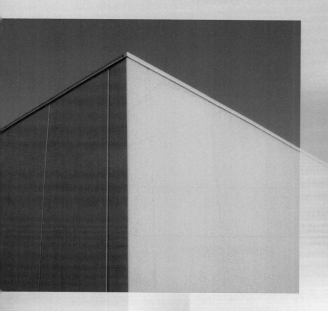

PRIOR LEARNING 1 Hang-glider

A hang-glider is a flying device which was initially designed to assist space capsules re-enter the atmosphere. Today, thousands of people all over the world have taken up hang-gliding for recreational purposes. The diagrams below show the image of a hang-glider and its mounting diagram.

a. What is the ratio of the dimensions of the diagram to the image of the hang-glider?

b. Complete the following proportions.

1) $\dfrac{m\,\overline{DE}}{m\,\overline{JK}} = \dfrac{m\,\overline{DF}}{\rule{1.5cm}{0.4cm}}$

2) $\dfrac{\rule{1.5cm}{0.4cm}}{m\,\overline{IJ}} = \dfrac{m\,\overline{AF}}{m\,\overline{GL}}$

c. Considering that angle 4 measures 65°, find the measure of the following angles. Justify your answers.

1) angle 1

2) angle 3

3) angle 5

4) angle 6

d. Determine:

1) the actual wingspan of the hang-glider

2) the scale used for the image

3) the area of the hang-glider

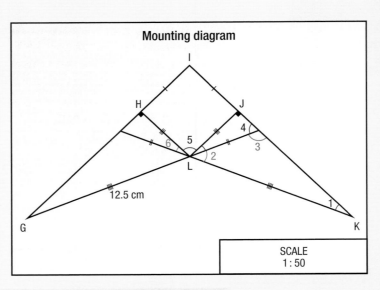

Image of a hang-glider

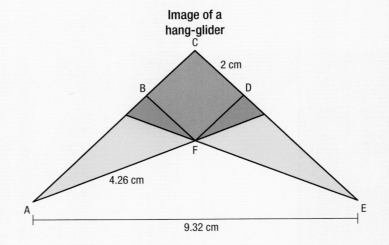

Mounting diagram

SCALE
1 : 50

An aircraft's wingspan is defined as the distance between wing tips. The shorter the wingspan of an airplane or glider, the easier it is to fly. The fastest airplane in the world, the X-43, has a wingspan of only 1.5 m.

knowledge summary

ANGLES

The sum of the measures of the interior angles of a triangle is 180°.

Two angles are **complementary** if the sum of their measures is 90°.

E.g. 1)

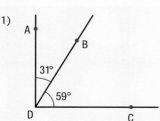

Angles ADB and BDC are complementary,
m∠ADB + m∠BDC = 31° + 59° = 90°.

2)

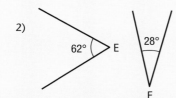

Angles E and F are complementary,
m∠E + m∠F = 62° + 28° = 90°.

Two angles are **supplementary** if the sum of their measures is 180°.

E.g. 1)

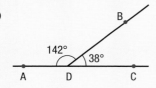

Angles ADB and BDC are supplementary,
m∠ADB + m∠BDC = 142° + 38° = 180°.

Angles E and F are supplementary,
m∠E + m∠F = 120° + 60° = 180°.

2)

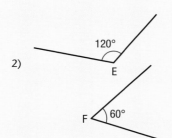

RATIOS

A **ratio** is a way of **comparing** two quantities or two magnitudes of the **same type** expressed in the **same units**, involving the concept of **division**. The two most common ways of writing a ratio or a rate are with a colon or a fraction bar. Thus, the ratio between a and b is written as

$a : b$ or $\frac{a}{b}$ where $b \neq 0$.

E.g. 1) The ratio between the length of segment BC and the length of segment AC can be written as 3:4 and is equivalent to $\frac{3}{4}$ and is equivalent to 0.75.

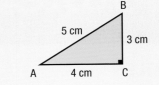

2) The ratio between the perimeter of triangle DEF and that of triangle ABC is:
- 18:12 or 3:2
- $\frac{18}{12}$ or $\frac{3}{2}$

This means that the perimeter of triangle DEF is 1.5 times greater than that of triangle ABC.

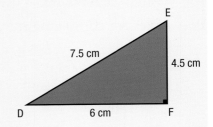

PROPORTION

A **proportion** is a statement of equality between two ratios.

If the ratio between a and b where $b \neq 0$ is equal to the ratio between c and d where $d \neq 0$, then $a : b = c : d$ or $\frac{a}{b} = \frac{c}{d}$ is a proportion.

A proportion consists of four terms. The first and fourth terms are called the **extremes**, and the second and third terms are called the **means**.

Means

$$a : b = c : d$$

Extremes

Extremes
$$\frac{a}{b} = \frac{c}{d}$$
Means

In a proportion the **product of the extremes** is equal to the **product of the means**.

E.g. Considering that the two shapes below are similar, $\frac{m\ \overline{CD}}{m\ \overline{GH}} = \frac{m\ \overline{AD}}{m\ \overline{EH}}$.

$$\frac{3}{x} = \frac{5}{6} \Rightarrow 5x = 18 \Rightarrow x = 3.6\ \text{cm}$$

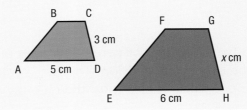

knowledge in action

1 Match each triangle in the table on the left with its corresponding similar triangle in the table on the right.

Triangle	Lengths of the sides of triangle ABC (cm)		
	m$\overline{AB}$	m$\overline{BC}$	m$\overline{CA}$
1	3	5	6
2	13.5	37.5	54
3	5.25	5.25	2.75
4	3.75	4.5	8

Triangle	Lengths of the sides of triangle DEF (cm)		
	m$\overline{DE}$	m$\overline{EF}$	m$\overline{FD}$
A	9	25	36
B	21	21	11
C	4.5	5.4	9.6
D	9	15	18

2 In each of the following proportions, find the missing term(s).

a) $\dfrac{3}{5} = \dfrac{14}{x}$
b) $6 : y = 4 : 3$
c) $\dfrac{z}{2.6} = \dfrac{4.5}{11.25}$
d) $24.5 : a = a : 2$

3 **ARTEMISION** The Artemision, an ancient Greek temple, consists of a set of 18-m columns that support a roof. These columns are arranged so as to form a rectangular base with a length of 130 m and a width of 68 m.

a) Determine the value of the following ratios:

1) $\dfrac{\text{length of the base of the temple}}{\text{width of the base of the temple}}$
2) $\dfrac{\text{width of the base of the temple}}{\text{height of the columns}}$

3) $\dfrac{\text{perimeter of the base of the temple}}{\text{height of the columns}}$

The temple of Artemis of Ephesus is one of the Seven Wonders of the Ancient World, which includes among others, the Hanging Gardens of Babylon and the Great Pyramids of Egypt. This temple is considered to be one of the first banks in the world.

b) The three illustrations below represent architectural models of the Artemision. Which of these three models is not drawn to scale? Justify your answer.

1)

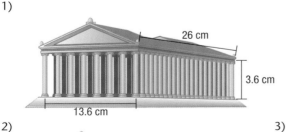

26 cm

3.6 cm

13.6 cm

2)

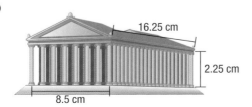

16.25 cm

2.25 cm

8.5 cm

3)

31.2 cm

4.82 cm

16.32 cm

4 Determine the measure of the complementary angle for each of the angles below.

a) 36°   b) 10.45°   c) 45°   d) 66.3°   e) 90°

5 On a map, two villages, **A** and **B**, are 3.5 cm apart. Determine the actual distance, in kilometres, between these two villages if the ratio between the distances on the map and the real distances is 1:200 000.

6 Consider the four trapezoids below.

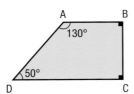 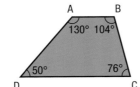 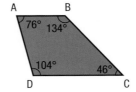 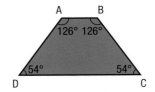

a) For each of these trapezoids, calculate $m\angle A + m\angle D$ and $m\angle B + m\angle C$. What do you notice?

b) Considering that one of the properties of a right triangle is that the two acute angles are complementary, do the following:

  1) Show that angles B and C in the adjacent trapezoid are supplementary.

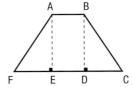

  2) Show that the consecutive angles in the adjacent parallelogram are supplementary.

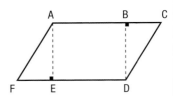

7 In the adjacent diagram, right Triangles ①, ②, ③, ④ and ⑤ are similar.

a) Determine the length of each side for each of these triangles.

a) Calculate the ratio $\dfrac{\text{length of the hypotenuse of Triangle ①}}{\text{length of the hypotenuse of Triangle ⑤}}$.

b) What is the total area of the figure comprised of Triangles ①, ②, ③, ④ and ⑤?

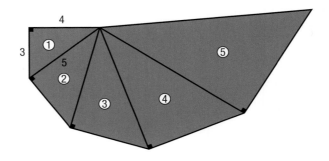

This section is related to LES 10.

PROBLEM   Water-skiing

A water-ski centre offers a variety of jump ramps each built with an angle of elevation of 12°. The more skilled the jumpers are, the longer the ramps they use. These ramps are in the form of triangular-based right prism. One of these ramps is illustrated below:

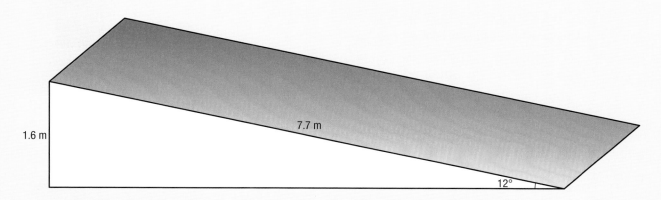

1.6 m

7.7 m

12°

A quality control employee at the ramp manufacturer must verify that the triangular bases of the ramps are indeed right triangles with a 12° inclination. The only tool available is a tape measure.

How can this employee confirm that the bases of these ramps are actually shaped as right triangles with a 12° inclination?

Water-skiing is one of the sport disciplines at the Pan American Games held every four years; these games bring together athletes from all nations of the Americas. The average speed of competitors in water-ski jumping events is 120 km/h.

ACTIVITY 1   Extension ladders

When forced to enter a burning building through a window, firefighters often use an extension ladder attached to the top of the fire truck.

Firefighters park the truck a certain distance away from the burning building and deploy the ladder at an angle of elevation of 50° from the horizon. Below are some examples.

a. What geometric statement allows you to state that triangle ABC, DEF and GHI are similar to each other?

b. Determine the length of the ladder in each of the adjacent examples.

c. 1) For each triangle, calculate the ratio

$$\frac{\text{length of the side opposite to the 50° angle}}{\text{length of the hypotenuse}}.$$

    2) Compare the results obtained.

    3) On your calculator, find the **SIN** key, for sine, and calculate the sine of 50°.

    4) What does the sine of an angle allow you to calculate?

d. 1) For each triangle, calculate the ratio

$$\frac{\text{length of the side adjacent to the 50° angle}}{\text{length of the hypotenuse}}.$$

    2) Compare the results obtained.

    3) On your calculator, find the **COS** key, for cosine, and calculate the cosine of 50°.

    4) What does the cosine of an angle allow you to calculate?

e. 1) For each triangle, calculate the ratio

$$\frac{\text{length of the side opposite to the 50° angle}}{\text{length of the side adjacent to the 50° angle}}.$$

    2) Compare the results obtained.

    3) On your calculator, find the **TAN** key, for tangent, and calculate the tangent of 50°.

    4) What does the tangent of an angle allow you to calculate?

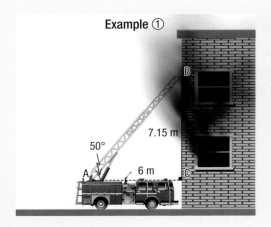

Example ②

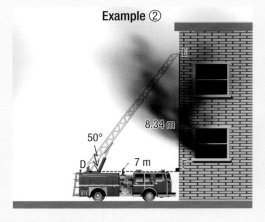

Example ①

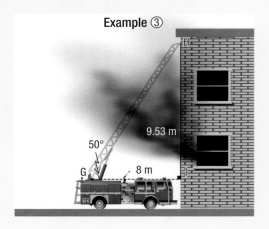

Example ③

Techno math

Dynamic geometry software allows you to draw right triangles and then manipulate them. By using the tools, LINE, PERPENDICULAR LINE, TRIANGLE, MARK AN ANGLE, ANGLE MEASUREMENT and DISTANCE, you can draw a right triangle on a Cartesian plane and display the lengths of its sides and the measures of its angles.

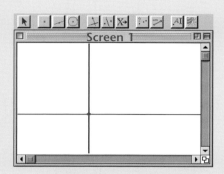

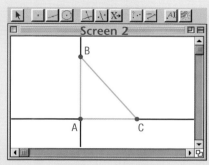

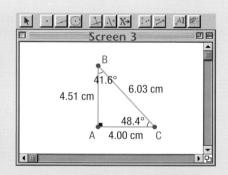

As you change the position of the vertices of the triangle, you will observe certain changes with respect to the measures of its angles and the lengths of its sides.

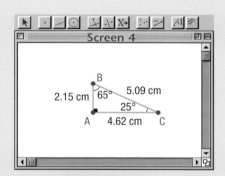

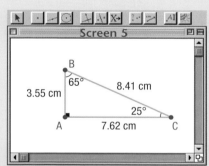

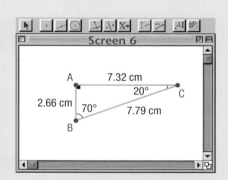

a. 1) Determine the ratio $\dfrac{\text{length of the side opposite to angle C}}{\text{length of the hypotenuse}}$ in the triangles displayed in Screens **3** through **6**.

2) Based on the above results, can you state that the greater the measure of angle C, the greater the ratio $\dfrac{\text{length of the side opposite to angle C}}{\text{length of the hypotenuse}}$?

b. 1) Referring to Screens **3** to **6**, determine the ratio $\dfrac{\text{length of the side adjacent to angle C}}{\text{length of the hypotenuse}}$.

2) Based on the above results, can you state that the greater the measure of angle C, the greater will be the ratio $\dfrac{\text{length of the side adjacent to angle C}}{\text{length of the hypotenuse}}$?

c. Using dynamic geometry software to move point B, determine the following:

1) What number does the ratio $\dfrac{\text{length of the side adjacent to angle C}}{\text{length of the hypotenuse}}$ approach as segment BC moves closer to 0°?

2) What number does the ratio $\dfrac{\text{length of the side opposite to angle C}}{\text{length of the hypotenuse}}$ approach as segment BC moves closer to 90°?

TRIGONOMETRIC RATIOS IN A RIGHT TRIANGLE

Trigonometry is the study of the relationship between the angles and the side lengths of a triangle.

A **trigonometric ratio** is a ratio of the lengths of two sides of a right triangle.

In a right triangle, the three main trigonometric ratios are:

$$\sin A = \frac{\text{length of the leg opposite to } \angle A}{\text{length of the hypotenuse}}$$

$$\cos A = \frac{\text{length of the leg adjacent to } \angle A}{\text{length of the hypotenuse}}$$

$$\tan A = \frac{\text{length of the leg opposite to } \angle A}{\text{length of the leg adjacent to } \angle A}$$

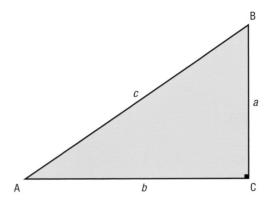

"Sin," "cos" and "tan" respectively signify sine, cosine and tangent.

E.g. In the adjacent triangle DEF:

$$\sin E = \frac{12}{20} = 0.6 \qquad \sin F = \frac{16}{20} = 0.8$$

$$\cos E = \frac{16}{20} = 0.8 \qquad \cos F = \frac{12}{20} = 0.6$$

$$\tan E = \frac{12}{16} = 0.75 \qquad \tan F = \frac{16}{12} \approx 1.\overline{3}$$

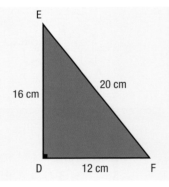

Solving a triangle means finding the lengths of its sides and the measures of its angles using some known data.

E.g. In the adjacent triangle GHI, note the following:

- Since the sum of the interior angles in a triangle is 180°, $m \angle H = 55°$.

- Since $\tan 35° = \frac{m \overline{GH}}{5}$, $m \overline{GH} = 5 \tan 35°$
≈ 3.5 cm.

- Since $\cos 35° = \frac{5}{m \overline{HI}}$, $m \overline{HI} = \frac{5}{\cos 35°}$
≈ 6.1 cm.

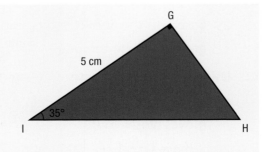

practice 5.1

1 Considering that R represents the vertex of one of the acute angles of a right triangle, complete the following table.

m ∠ R	sin R	cos R	tan R
15°			
30°			
45°			
60°			
75°			

2 For each case, determine the value of the sine, the cosine and the tangent of angle G.

a)

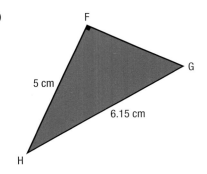

5 cm
6.15 cm

b)

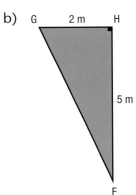

2 m
5 m

c)

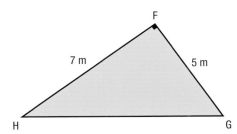

7 m
5 m

d)

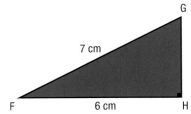

7 cm
6 cm

e)

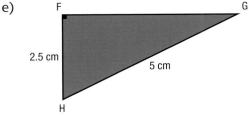

2.5 cm
5 cm

f)

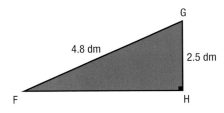

4.8 dm
2.5 dm

3 Solve the following triangles.

a)
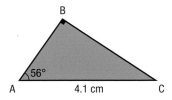

b)
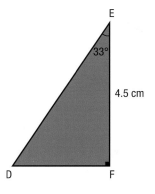

c)
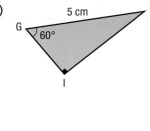

c)
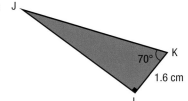

d)
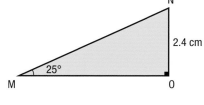

f)
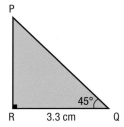

4 The adjacent diagram shows three parallel boards resting against a wall.

a) At what height does Board ① touch the wall?

b) What is the length of Board ②?

c) At what distance from the wall does the base of Board ③ rest?

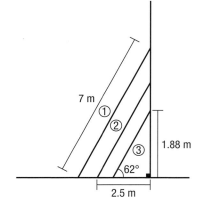

5 Using the adjacent right triangle, indicate whether the following statements are true or false. If the statement is true, explain why. If the statement is false, provide a counter-example.

a) $\dfrac{\sin S}{\sin T} = \dfrac{\cos S}{\cos T}$

b) $\tan S = \dfrac{\cos T}{\cos S}$

c) $\sin S + \cos S = \tan S$

d) $\sin R = 1$

e) The cosine of an angle cannot be greater than 1.

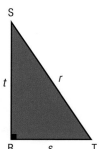

6 The front view of a martini glass is represented as two isosceles right triangles. What is the distance between the extremities C and E of this martini glass?

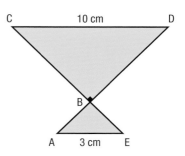

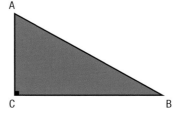

7 Considering the adjacent diagram, complete the following statements using the symbols $<$, $>$ or $=$.

a) If sin A $>$ sin B, then m∠A ___ m∠B.

b) sin A ___ cos B

c) sin C ___ sin A + sin B

d) tan A ___ $\dfrac{1}{\tan B}$

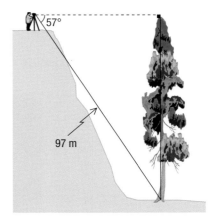

8 A logging company must determine the height of three, 400-year-old Douglas fir trees located in a British Columbia forest. Consider the information below:

Tree ①   Tree ②   Tree ③

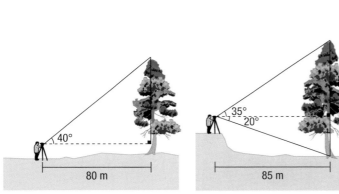

Taking into account the fact that the device used to determine the angle from the ground to one of the ends of the tree is 1.5 m above the ground, determine the height of each tree.

The *Pseudotsuga menziesii* or Douglas fir is a rapid-growth species with a life expectancy of 400 to 500 years. Among other things, this tree is used for reforestation and can reach heights of 50 to 80 m with trunks up to 2 m in diameter.

9 a) Complete the following table.

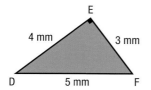

	sine	cosine	$\dfrac{\text{sine}}{\text{cosine}}$	tangent
Angle D	$\dfrac{3}{5}$	▢	▢	▢
Angle F	▢	▢	▢	▢

b) Based on the results obtained in **a)**, what conjectures can you formulate?

c) Using the adjacent triangle ABC, prove that these conjectures are true for all right triangles.

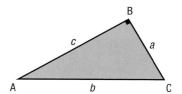

10 As shown in the adjacent diagram, a camper observes a boat come loose from its anchored location and watches it move with the current up to a beaver dam.

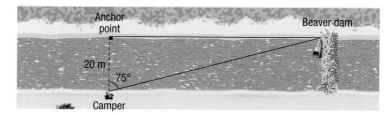

The camper estimates that the boat travelled at least 75 m. Is this correct? Justify your answer.

11 Leah and Mara decide to take different routes to cross a wooded area.

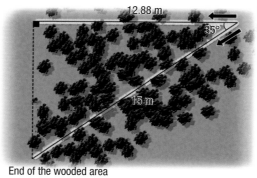

Referring to the data in the above diagram, answer the following questions:

a) How many seconds will it take for Leah to get to point B?

b) How many seconds will it take for Mara to get to point B?

c) At what speed should Leah walk in order to arrive at the end of the wooded area at the same time as Mara?

12 Some camper trailers have an awning that provides shelter from rain and sun. These awnings are often installed at an angle to prevent water from accumulating.

The picture below shows an awning that has been set-up. If point A is located 2.8 m above the ground, what is the height of the post represented by segment DG?

13 It is possible to calculate the distance separating the Sun from a star by observing the position of this star every six months. Referring to the illustration below, determine the distance between the Sun and the star 61 Cygni.

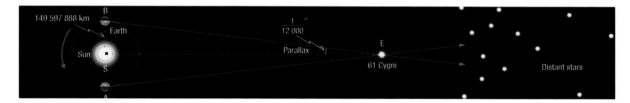

14 **PARALLELS** To determine our position on the Earth's surface, a system of imaginary lines that crisscross the globe are used; these are called *meridians* and *parallels*. The latter are circles parallel to the Equator; they are numbered according to the angle formed with the centre of the Earth. The meridians are semi-circles that link the two poles. Considering that the Equator has a circumference of 40 075 km, angle AOC measures 20° and angle BOC measures 60°, determine:

a) the circumference of the 20th north parallel

b) the length of the segment between two cities **A** and **B** on the same meridian, one located on the 20th north parallel, the other on the 60th south parallel

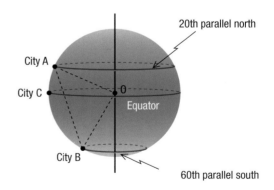

A large part of the border between Canada and the United States straddles the 49th north parallel. In fact, many borders follow, in whole or in part, meridians and parallels. To this day, the Canadian-American border is the longest border in the world.

15 At a summer camp, an orientation game using maps and compasses is organized. Each team must follow the instructions below in order to complete the course.

1. From the starting point A, walk east to point B.
2. Walk 56.57 m in a southwesterly direction to point C.
3. Moving in a northwesterly direction, walk to point D, 30 m west of the starting point.
4. Turn in a north-northeasterly direction toward point E, which is north of point A.
5. Go back to the starting point.

Determine the total distance required to complete the course.

A compass is an instrument that uses magnetized needle to point the north. The term "magnetic compass" is primarily associated with terrestrial orientation. In the field of aerial and marine navigation, aircraft and ships are steered using a device similarly called a "compass."

16 Luka starts a chronometer as an F-18 Hornet flies directly over head in an easterly direction. He then takes note of its position 8 s later. The illustration below represents this situation. Considering that this type of aircraft cannot fly at an altitude higher than 18 000 m, what is maximum speed of this aircraft?

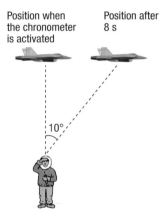

Position when the chronometer is activated

Position after 8 s

10°

The Mach number corresponds to a multiple of the speed of sound. An airplane flying at Mach 3 flies at a speed three times the speed of sound. On the other hand, the F-18 Hornet flies at supersonic speed.

This section is related to LES 10.

 PROBLEM Radar

A radar antenna is a device that calculates the distance, position and the speed of various objects such as airplanes or boats. To do this, the antenna produces airwaves that, when they meet an obstruction, are reflected back and captured by the same antenna.

A radar antenna is installed at point A and covers the region defined by segments AC and AD. The antenna is then moved to point B located 5 km away from point A as shown in the diagram below.

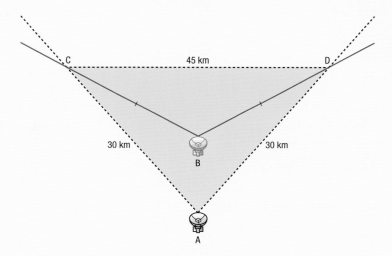

 What is the measure of the angle of emission, CBD, of this antenna if it must cover the region defined by segments BC and BD?

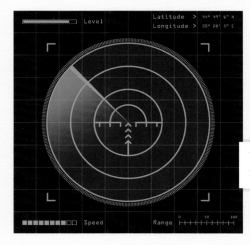

The word "radar" is an acronym for *radio detection and ranging*.

An individual who wishes to take a pilot training course must undergo a medical examination including a particular type of eye examination to verifying that the angular deflection of the right eye is strictly less than 3°. Angular deflection is defined as the angle formed by the actual position of the object being observed, the eye and the perceived position of the object. Below are the result obtained by three candidates:

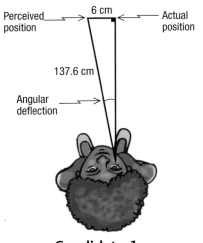

Candidate 1

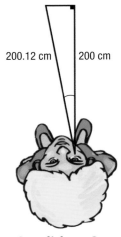

Candidate 2

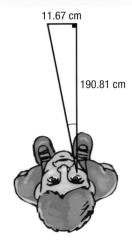

Candidate 3

a. Referring to Candidate **1**, determine:

1) the sine of the angular deflection of her eye

2) the angular deflection of her eye using the adjacent table of values

3) whether or not the candidate can take a pilot training course

Measure of the angle (°)	Sine
1	≈ 0.017 4
1.5	≈ 0.026 2
2	≈ 0.034 9
2.5	≈ 0.043 6
3	≈ 0.052 3
3.5	≈ 0.061 0

b. Carefully examining the adjacent calculator displays, determine:

1) the calculation performed by the $\sin^{-1}$ key

2) the measure of the angle whose sine is 0.866 025 4038

```
sin(14)
           .2419218956
sin⁻¹(Ans)
               14
sin(30)
               .5
sin⁻¹(0.5)
               30
sin⁻¹(0.866025403
8)
               60
sin⁻¹(0.071067812
)
               45
```

c. Referring to Candidate **2**, determine:

1) the cosine of the angular deflection of her eye

2) the angular deflection of her eye using the adjacent table of values

3) whether or not the candidate can take a pilot training course

Measure of the angle (°)	Cosine
1	≈ 0.999 8
1.5	≈ 0.999 7
2	≈ 0.999 4
2.5	≈ 0.999 0
3	≈ 0.998 6
3.5	≈ 0.998 1

d. Carefully examining the calculator display below, determine:
1) the calculation performed by the cos⁻¹ key
2) the measure of the angle whose cosine is 0.258 819 0451

```
cos(9)
            .9876883406
cos⁻¹(Ans)
                      9
cos(60)
                     .5
cos⁻¹(0.5)
                     60
cos⁻¹(0.939692620
8
                     20
cos⁻¹(0.258819045
1
                     75
```

e. Referring to Candidate **3**, determine:
1) the tangent of the angular deflection of his eye
2) the angular deflection of his eye using the adjacent table of values
3) whether or not the candidate can take a pilot training course

Measure of the angle (°)	Tangent
1	≈ 0.0174 6
1.5	≈ 0.0261 9
2	≈ 0.0349 2
2.5	≈ 0.0436 6
3	≈ 0.0524 1
3.5	≈ 0.0611 6

f. Carefully examining the adjacent calculator displays, determine:
1) the calculation performed by the tan⁻¹ key
2) the measure of the angle whose tangent is 0

```
tan(37)
            .7535540501
tan⁻¹(Ans)
                     37
tan(45)
                      1
tan⁻¹(1)
                     45
tan⁻¹(1.962610506
)
            63.00000001
tan⁻¹(0)
                      0
```

g. 1) Using a calculator, determine the measure of angle A in each of the following triangles.

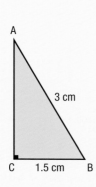

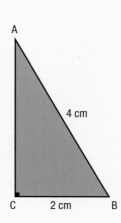

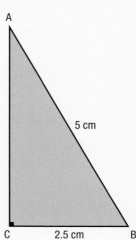

2) Taking the above answers into consideration, what conjecture can you formulate?

Techno math

Dynamic geometry software allows you to explore and verify certain relations. By using the tools REGULAR POLYGON, BISECTOR, TRIANGLE, MARK AND TRIANGLE, ANGLE MEASUREMENT and DISTANCE OR LENGTH, you can draw a right triangle where one of the angles measures 30°.

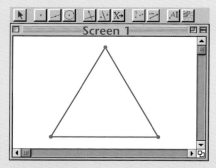
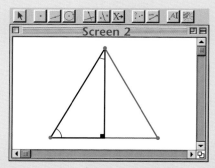
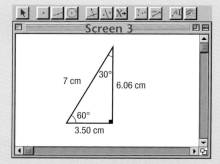

By moving one of the vertices of the triangle, you can observe certain changes related to the dimensions of the triangle.

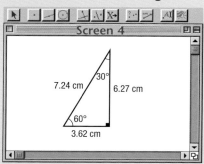
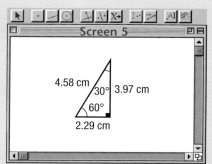
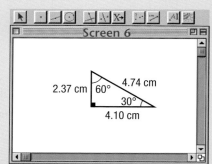

a. What do the right triangles in Screens **3** to **6** have in common?

b. 1) Complete the table below.

Trigonometric ratio	Screen 3	Screen 4	Screen 5	Screen 6
length of side opposite to 30° angle / length of hypotenuse				
length of side adjacent to 30° angle / length of hypotenuse				
length of side opposite to 60° angle / length of hypotenuse				
length of side adjacent to 60° angle / length of hypotenuse				

2) Based on the results obtained in the table above, what conjecture can you formulate regarding a right triangle where one of the angles measures 30°?

c. 1) Using dynamic geometry software, draw a right triangle.
2) Move the vertices of this triangle and determine how many angle measures result in a sine of 0.5.

d. Verify whether the conjecture formed in **b. 2)** is true for three other right triangles where one of the angles measures 30°.

knowledge 5.2

LENGTHS IN A 30° RIGHT TRIANGLE

In a right triangle, the length of the side opposite a 30° angle is equal to half the length of the hypotenuse.

E.g. In the adjacent right triangle, the following is true:

• If m $\overline{BC}$ = 10 cm, then m $\overline{AB}$ = 5 cm.

• If m $\overline{AB}$ = 42 mm, then m $\overline{BC}$= 84 mm.

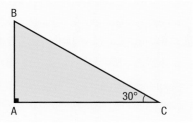

SOLVING A RIGHT TRIANGLE

Considering that sin, cos and tan allow you to calculate the trigonometric ratios of given angle measurements; arcsin, arccos and arctan allow you to do the inverse, in other words, calculate angle measurements based on the values of the corresponding ratios.

"Arcsin," "arccos" and "arctan" are respectively associated with the functions arcsine, arccosine and arctangent.

E.g. In the right triangle below:

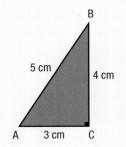
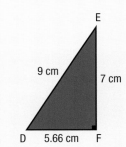
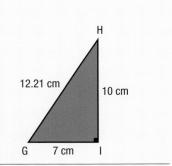

You have:	You have:	You have:
$\sin A = \frac{4}{5}$ and arcsin $\frac{4}{5} \approx 53.13°$	$\cos E = \frac{7}{9}$ and arccos $\frac{7}{9} \approx 38.94°$	$\tan G = \frac{10}{7}$ and arctan $\frac{10}{7} \approx 55.01°$
You can deduce that:	You can deduce that:	You can deduce that:
m∠A ≈ 53.13°	m∠E ≈ 38.94°	m∠G ≈ 55.01°
m∠B ≈ 36.87°	m∠D ≈ 51.06°	m∠H ≈ 34.99°

In summary, you can solve for a right triangle if you know:

• the measurements of one acute angle and one side, using sin, cos and tan

• the length of two sides, using arcsin, arccos and arctan

> Arcsin, arccos and arctan can also be written as $\sin^{-1}$, $\cos^{-1}$ and $\tan^{-1}$.

practice 5.2

1 For each of the following, determine the measure of the angle associated with the expression given.

a) arcsin 0.17   b) arcsin 0.5   c) arcsin 0.87   d) arccos 0.91

e) arccos 0.71   f) arccos 0.26   g) arctan 0.7   h) arctan 1

2 Refer to the adjacent right triangle and determine which of the following equalities are true.

a) $\sin^{-1} \frac{g}{e} = \cos^{-1} \frac{g}{e}$

b) $\cos^{-1} \frac{f}{e} = \sin^{-1} \frac{g}{e}$

c) $\tan^{-1} \frac{f}{e} = \tan^{-1} \frac{e}{f}$

d) $\sin^{-1} \frac{f}{e} = \tan^{-1} \frac{f}{g}$

e) $\cos^{-1} \frac{g}{e} = \sin^{-1} \frac{e}{f}$

f) $\tan^{-1} \frac{g}{f} = \cos^{-1} \frac{f}{e}$

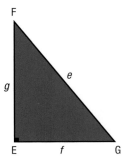

3 Consider a right triangle ABC whose right angle is located at vertex B. Determine the measure of angle A if:

a) sin A = 0.34   b) sin A = 0.96   c) cos A = 0.99

d) cos A = 0.12   e) tan A = 0.40   f) tan A = 28.63

4 The table below provides some information on several right triangles ABC whose right angle is located at vertex C. Complete the table.

	Length of side adjacent to angle A (cm)	Length of side opposite to angle A (cm)	Length of hypotenuse (cm)	Measure of angle A (°)	Measure of angle B (°)
Triangle ①	12		35		
Triangle ②		30	60		
Triangle ③	11.4	15.5			
Triangle ④	45.76			30	
Triangle ⑤			0.45		27
Triangle ⑥	34.5	46	57.5		

5 A tree that is 20 m tall projects a shadow 16 m long. What is the inclination of the Sun's rays when this shadow is projected?

6 Solve the triangles below.

a)
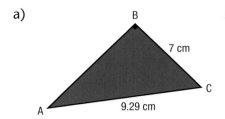

b)
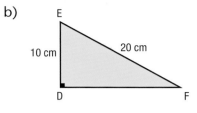

c)
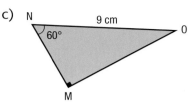

d)
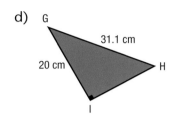

e)
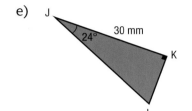

f)
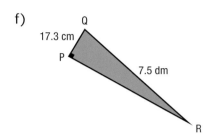

7 The adjacent diagram illustrates the position of a boat heading toward Roché Percé. What angle should the helmsman give to the boat's trajectory in order to sail past:

a) west of Roché Percé?

b) east of Roché Percé?

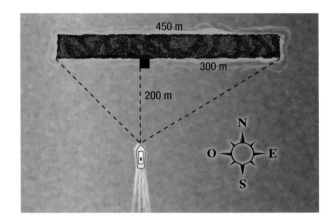

It is believed that Roché Percé once had four arches, but only one 30 m wide remains today.

8 As illustrated in the diagram below, a ski lift is designed to bring skiers to the top of a mountain. What is the measure of the angle of inclination of this ski lift?

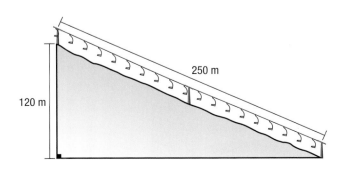

9 **THE CHEOPS PYRAMID** The Cheops Pyramid is a right regular pyramid that is 137 m tall and has a square base measuring 230 m on each side.

a) What is the measure of the angle formed by the base of the pyramid and one of its faces?

b) What is the measure of the angle formed by a diagonal of the base of the pyramid and an edge shared by two lateral faces of this pyramid?

c) What are the measures of the three angles of each lateral face of the pyramid?

For over 4500 years, the Cheops Pyramid has been the world's most studied monument.

10 The diagram below shows the oscillating motion of a pendulum. If the distance between the two extreme positions, A and B, reached by the pendulum is 10 cm, what is the length of the pendulum's string?

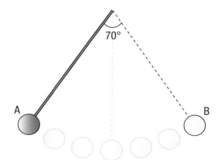

A   70°   B

11 Aisha observes the top of the Statue of Liberty from a spot located 46.5 m away from its base; her gaze follows an angle of elevation of 60°. Considering that Aisha's eyes are at a height of 1.62 m above the ground, how tall is of the Statue of Liberty?

In 1866, The Statue of Liberty was given, by the French, to the United States in commemoration of one hundred years of American independence and as a token of friendship between the two nations.

12 The diagram below shows the elevation of a dam shaped like an isosceles trapezoid. Metal braces placed along the diagonals of the trapezoid are used to strengthen its structure. Based on the information provided, what is the total length of the braces?

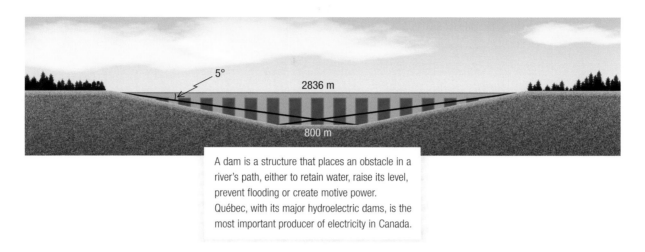

A dam is a structure that places an obstacle in a river's path, either to retain water, raise its level, prevent flooding or create motive power.
Québec, with its major hydroelectric dams, is the most important producer of electricity in Canada.

13 GRAND CANYON An individual observes the shores of the river at the bottom of the Grand Canyon. At this location, the Canyon has a depth of 1300 m and a width of 5.5 km. The diagram below illustrates this situation.

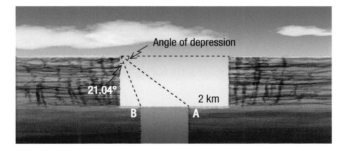

a) What is the width of the river at this location?

b) What is the measure of the angle of depression viewed by this person from the furthest shore?

c) What would be the measures of the angles of depression if this same individual observed the shores from the other side of the Canyon?

The Grand Canyon, located in Arizona, extends over approximately 450 km. It contains some of the most ancient rocks on Earth; some of these rocks, found at the bottom of the gorges, are 1.7 billion years old.

14 At 1:28 p.m., two airplanes are facing each other and are preparing to land on the same runway. The diagram below provides some information noted by the air traffic controller at this precise moment.

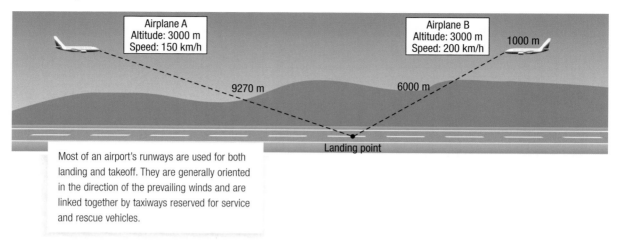

Most of an airport's runways are used for both landing and takeoff. They are generally oriented in the direction of the prevailing winds and are linked together by taxiways reserved for service and rescue vehicles.

a) What is the distance separating the two airplanes?

b) What is the measure of the angle of descent that the air traffic controller must indicate to the pilots if both must land at the same spot on the runway?

c) What will be the altitude of Airplane **A** at the moment that Airplane **B** touches the runway?

15 In 1898, two American kite flyers established a world record. Henry Helm and A.E. Sweetland succeeded in flying their kites at a height of 3801 m.

a) If they used a tether line 8993.93 m long, what was the angle of elevation defined by the horizon and the tether line?

b) If they had wanted to increase the height of the kite by 1 m while maintaining the same angle of elevation, what length should they have added to the tether line?

Kites have played an important role in scientific research and meteorology. Some examples include Benjamin Franklin's experiments with electricity, the first flight by the Wright brothers and Alexander Graham Bell's six-sided kite.

16 The windmill shown in the adjacent diagram has four identical sails in the shape of isosceles triangles. At what height from the ground will point A be situated if it rotates:

a) 235° counter-clockwise?

b) 50° clockwise?

c) 455°?

d) -690°?

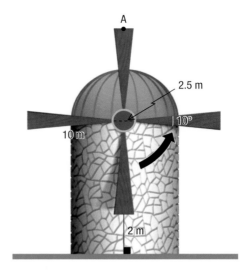

17 **ERATOSTHENES** Eratosthenes, an ancient Greek thinker, noticed that everyday at the same time, the Sun's rays would shine perpendicularly onto the town of Syene yet slant at an angle on the city of Alexandria. By measuring the length of the shadow of a pillar in Alexandria and the distance between the two cities, he was able to determine the circumference of the Earth. The diagram below represents this situation.

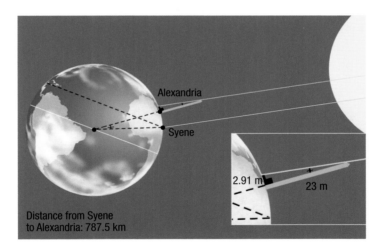

Alexandria

Syene

2.91 m   23 m

Distance from Syene to Alexandria: 787.5 km

a) What is the measure of the angle formed by the Sun's rays and the pillar in Alexandria?

b) What is the measure of the central angle shown in the diagram?

c) According to Eratosthenes' method, what is the circumference of the Earth?

Eratosthenes is said to have estimated the distance between the cities of Syene and Alexandria on the basis of the number of days it took to get from one to the other while travelling by camel. His estimate was 785 km, which was very close to the actual distance.

ERATOSTHENES

This section is related to LES 11.

PROBLEM   Hitting the best serve

In tennis, when playing doubles, the person serving can hit the ball from anywhere between points A and B outside of the court. The following illustrations show Thomas' options depending on whether he serves from point A or point B.

Example 1

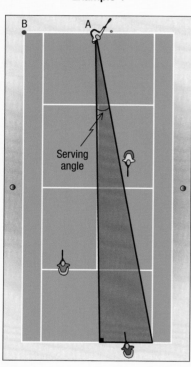

Serving angle

Example 2

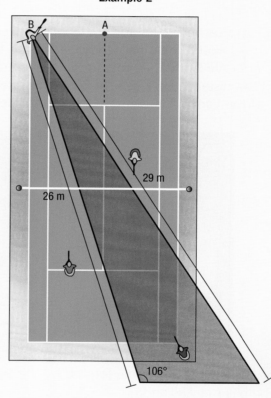

29 m

26 m

106°

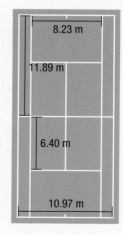

8.23 m

11.89 m

6.40 m

10.97 m

As Thomas moves closer to point B and prepares to serve, the measure of his serving angle increases.

In relation to his serving angle, how many degrees does Thomas gain by placing himself at point B rather than point A?

ACTIVITY 1 The sine law

Michelle observes the measurements of the sides and angles of triangles ABC and DEF shown below:

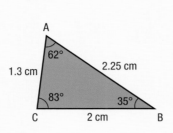

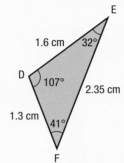

OBSERVATION

In a triangle, the largest angle is opposite to the longest side and the smallest angle is opposite the shortest side.

CONJECTURE

I therefore deduce that the lengths of the sides of a triangle are proportional to the measures of the angles opposite to those sides.

a. Referring to triangles ABC and DEF, verify that Michelle's observation is accurate.

b. For each triangle, Michelle represents her conjecture using proportions.

Triangle ABC

$$\frac{2.25 \text{ cm}}{83°} = \frac{2 \text{ cm}}{62°} = \frac{1.3 \text{ cm}}{35°}$$

Triangle DEF

$$\frac{2.35 \text{ cm}}{107°} = \frac{1.3 \text{ cm}}{32°} = \frac{1.61 \text{ cm}}{41°}$$

Check if these equalities are true.

c. Is Michelle's conjecture true? Justify your answer.

Joseph observes the measurement of the sides and angles of the same triangles ABC and DEF.

OBSERVATION

In a triangle, the sine of the largest angle is largest and the sine of the smallest angle is smallest.

CONJECTURE

I therefore deduce that the side lengths of a triangle are proportional to the sine of the angles opposite to those sides.

d. Referring to triangles ABC and DEF, verify that Joseph's observation is accurate.

e. For each triangle, Joseph represents his conjecture using proportions.

<table>
<tr><td align="center">**Triangle ABC**</td><td align="center">**Triangle DEF**</td></tr>
<tr><td align="center">$\dfrac{2.25 \text{ cm}}{\sin 83°} = \dfrac{2 \text{ cm}}{\sin 62°} = \dfrac{1.3 \text{ cm}}{\sin 35°}$</td><td align="center">$\dfrac{2.35 \text{ cm}}{\sin 107°} = \dfrac{1.3 \text{ cm}}{\sin 32°} = \dfrac{1.61 \text{ cm}}{\sin 41°}$</td></tr>
</table>

Check if these equalities are true.

f. Is Joseph's conjecture true? Justify your answer.

John and Audrey make the following statements.

JOHN'S OBSERVATION   AUDREY'S OBSERVATION

I noticed that the more an angle's measurement increases, so does the value of its sine.

I noticed that the sines of two supplementary angles are equal.

g. Calculate the value of the following expressions.
1) sin 15° and sin 165°
2) sin 30° and sin 170°
3) sin 45° and sin 135°
4) sin 60° and sin 129°
5) sin 85° and sin 110°

h. Based on the results found in **g.**, decide the following:
1) Is John's statement correct? Explain your answer.
2) Is Audrey's statement correct? Explain your answer.

ACTIVITY 2 A little bit of area

Heron of Alexandria discovered a special way to calculate the area A of any triangle. He stated the following formula:

$$A = \sqrt{p(p - a)(p - b)(p - c)}$$

where p represents the half-perimeter of the triangle and
a, b and c represent the lengths of the three sides of the triangle.

Heron of Alexandria, 1st century engineer, mechanic and Greek mathematician.

a. 1) Calculate the area of the adjacent triangle using the formula $A = \dfrac{\text{base} \times \text{height}}{2}$.

2) Calculate the area of this triangle using Hero's formula.

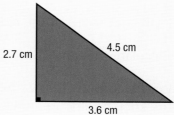

b. 1) Calculate the area of the adjacent triangle using the formula $A = \dfrac{\text{base} \times \text{height}}{2}$.

2) Calculate the area of this triangle using Hero's formula.

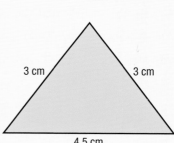

c. 1) Is it possible to calculate the area of the adjacent triangle using the formula $A = \dfrac{\text{base} \times \text{height}}{2}$?

2) Calculate the area of this triangle using Hero's formula.

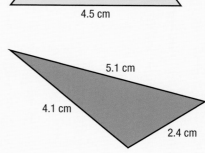

d. Identify an advantage for using Hero's formula instead of the formula $A = \dfrac{\text{base} \times \text{height}}{2}$.

Triangle ABC is shown below. Height *h* has been drawn from vertex B. The following reasoning allows you to calculate the area of triangle ABC if the lengths of sides AC and AB and the measure of angle A are known:

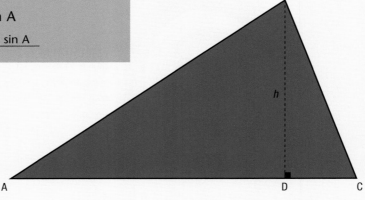

1) $A = \dfrac{m\,\overline{AC} \times m\,\overline{BD}}{2}$

2) Triangle ABD is a right triangle whose right angle is located at vertex D.

3) $\sin A = \dfrac{m\,\overline{BD}}{m\,\overline{AB}}$

4) $m\,\overline{BD} = m\,\overline{AB} \times \sin A$

5) $A = \dfrac{m\,\overline{AC} \times m\,\overline{AB} \times \sin A}{2}$

e. Justify each step of the reasoning shown above.

f. Calculate the area of triangle ABC if $m\,\overline{AC}$ = 15 cm, $m\,\overline{AB}$ = 12 cm and $m\angle A$ = 34°.

g. Write a formula that would allow you to calculate the area of triangle DEF shown below if the known measurements are:

1) $m\,\overline{DE}$, $m\,\overline{DF}$ and $m\angle D$

2) $m\,\overline{DE}$, $m\,\overline{EF}$ and $m\angle E$

3) $m\,\overline{EF}$, $m\,\overline{DF}$ and $m\angle F$

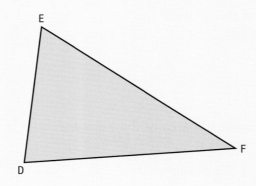

h. State the necessary conditions for using reasoning similar to the one expressed above.

Techno math

A graphing calculator allows you to design and use programs in order to generate quick calculations.

This screen allows you to execute, modify or create a new program.

Screen 1

```
EXEC EDIT NEW
1:Create New
```

Screen 2

```
PROGRAM
Name=HERO
```

This screen allows you to select certain programming instructions. For example, the "Prompt" instruction allows you to input a value for a given variable, and the "Disp" instruction allows you to display characters on the screen.

Screen 3

```
CTL I/O EXEC
1:Input
2:Prompt
3:Disp
4:DispGraph
5:DispTable
6:Output(
7↓getKey
```

This screen shows the commands of a program that allows you to calculate the area of any triangle where the lengths of the sides correspond to A, B and C, and whose half-perimeter corresponds to P.

Screen 4

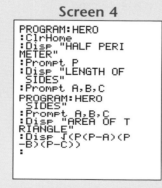

```
PROGRAM:HERO
:ClrHome
:Disp "HALF PERI
METER"
:Prompt P
:Disp "LENGTH OF
 SIDES"
:Prompt A,B,C
PROGRAM:HERO
 SIDES"
:Prompt A,B,C
:Disp "AREA OF T
RIANGLE"
:Disp √(P(P-A)(P
-B)(P-C))
:
```

Screen 5

```
EXEC EDIT NEW
1:HERO
```

Screen 6

```
prgmHERO
```

This screen shows the program being executed based on the measurements entered by the user.

Screen 7

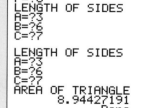

```
HALF PERIMETER
P=?8
LENGTH OF SIDES
A=?3
B=?6
C=?7

LENGTH OF SIDES
A=?3
B=?6
C=?7
AREA OF TRIANGLE
      8.94427191
           Done
```

a. Based on Screen **7**, answer the following:
1) What is the length of each side of the triangle?
2) What is the half-perimeter of the triangle?
3) What is the perimeter of the triangle?

b. Using a graphing calculator, determine the area of a triangle whose sides lengths are:
1) 5 cm, 7 cm and 8 cm   2) 2.2 dm, 4.4 dm and 5.5 dm   3) 390 m, 865 m and 490 m

c. Using a graphing calculator, modify the program shown in Screen **4** so that the user only needs to enter the lengths of the three sides of the triangle.

knowledge 5.3

RELATIONS IN ANY TRIANGLE

The sine law

It is possible to solve any triangle if you know the measurements of an angle, its opposite side and another side or angle of this triangle.

The length of the sides of a triangle are proportional to the sine of the angles opposite these sides. In the adjacent triangle:

$$\frac{a}{\sin A} = \frac{b}{\sin B} = \frac{c}{\sin C}$$

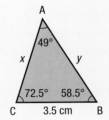

E.g. In the adjacent triangle:

$$\frac{x}{\sin 58.5°} = \frac{3.5}{\sin 49°} = \frac{y}{\sin 72.5°}$$

- Thus, $x = \dfrac{\sin 58.5° \times 3.5}{\sin 49°}$, $x \approx 3.95$ cm.

- Thus, $y = \dfrac{\sin 72.5° \times 3.5}{\sin 49°}$, $y \approx 4.42$ cm.

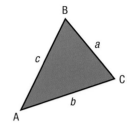

Hero's formula

When you know the length of the three sides, the area of the triangle is calculated using the formula $A = \sqrt{p(p-a)(p-b)(p-c)}$ where p represents the half-perimeter of the triangle: $p = \dfrac{a+b+c}{2}$.

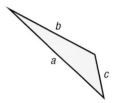

E.g. In the adjacent triangle:

$$p = \frac{1.5 + 2.5 + 3.5}{2} = 3.75 \text{ cm}$$

$$A = \sqrt{3.75(3.75 - 1.5)(3.75 - 2.5)(3.75 - 3.5)}$$

$$A \approx 1.62 \text{ cm}^2$$

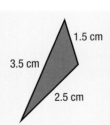

Trigonometric formula

It is possible to calculate the area of a triangle if you know the length of two sides and the measure of the contained angle.

The area of the triangle is calculated using the formula $A = \dfrac{a \times b \times \sin C}{2}$.

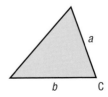

E.g. In the adjacent triangle:

$$A = \frac{3 \times 4 \times \sin 30°}{2}$$

$$A = 3 \text{ cm}^2$$

practice 5·3

1 For each of the following proportions, determine the value of *x*.

a) $\dfrac{35}{\sin 50°} = \dfrac{x}{\sin 10°}$

b) $\dfrac{x}{\sin 65°} = \dfrac{17.5}{\sin 15°}$

c) $\dfrac{98.3}{\sin 134°} = \dfrac{x}{\sin 30°}$

d) $\dfrac{20}{\sin x°} = \dfrac{75}{\sin 83°}$

e) $\dfrac{76}{\sin 76°} = \dfrac{34}{\sin x°}$

f) $\dfrac{90}{\sin 25°} = \dfrac{160}{\sin x°}$

2 For each of the triangles below, determine the missing measurement.

a)
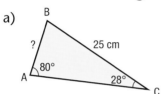

b)
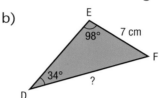

c)
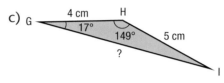

d)
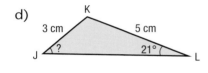

e)
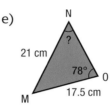

f)
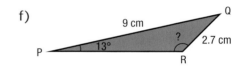

g)
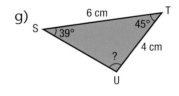

h)
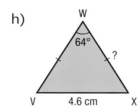

i)
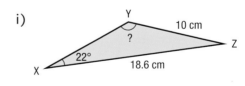

3 Refer to the adjacent triangle ABC in order to complete the table:

	m $\overline{BC}$	m ∠C	m ∠B	Area of △ABC
a)	4.8 cm			
b)	10 cm			
c)	5 cm			

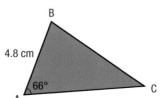

4 Consider the following calculations:

```
sin(30)
        .5
cos(30)
   .8660254038
tan(30)
   .5773502692
```

```
sin(45)
   .7071067812
cos(45)
   .7071067812
tan(45)
        1
```

```
sin(150)
        .5
cos(150)
   -.8660254038
tan(150)
   -.5773502692
```

```
sin(135)
   .7071067812
cos(135)
   -.7071067812
tan(135)
        -1
```

a) Verify whether the following equalities are true.

1) $\sin 30° = \sin (180° - 30°) = \sin 150°$

2) $\cos 45° = \cos (180° - 45°) = \cos 135°$

3) $\dfrac{\sin 150°}{\cos 150°} = \tan 150°$

b) What conjecture can you formulate about:

1) the sine of two supplementary angles?

2) the cosine of two supplementary angles?

3) the tangent of two supplementary angles?

5 Indicate whether the following equalities are true or false.

a) $\sin (30° + 10°) = \sin 30° + \sin 10°$

b) $\sin \left(\dfrac{30}{10}\right)° = \dfrac{\sin 30°}{\sin 10°}$

c) $\sin (30 \times 10)° = \sin 30° \times \sin 10°$

6 Calculate the area of each of the following triangles.

a)
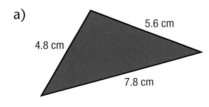
5.6 cm
4.8 cm
7.8 cm

b)
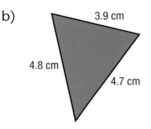
3.9 cm
4.8 cm
4.7 cm

c)
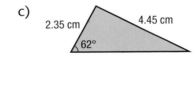
2.35 cm
4.45 cm
62°

7 The adjacent diagram shows a circle inscribed in an isosceles triangle. The centre of the circle corresponds to the intersection of the angle bisectors of the triangle.

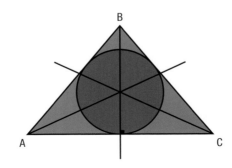

If m $\overline{AB}$ = m $\overline{BC}$ = 12 cm and m∠ABC = 80°, determine the area of the blue region.

8 Determine the area of each of the shapes below.

a)
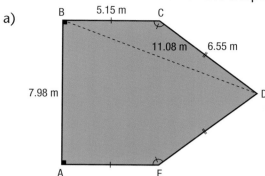

b)
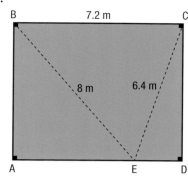

9 Determine the area of each of the following triangles.

a)
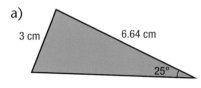

b)
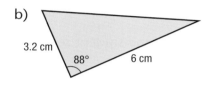

c)
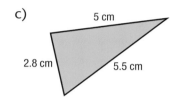

10 The length of the sides of a triangular field are 200 m, 175 m and 185 m. What would it cost to sod this field if sod costs $7/m²?

11 A carpenter uses many different tools including an accordion extension riveter similar to that shown in the adjacent picture.

The length of this accordion can vary from 4 cm to 40 cm. When the length of the accordion is 4 cm, angle A measures 10°. What is the measure of angle A when the accordion is extended to its maximum length?

12 While taking a walk in the woods, two campers use a compass and a pedometer. They walk 1000 paces in a southeast direction; they then turn south-southwest, walk 1500 paces and then stop. If each pace is 60 cm, what distance have they travelled from their starting point?

Known to the Chinese for a long time, it wasn't until about 1600 BCE that the use of compasses became popular in Europe. The four cardinal points are north, south, east and west; the names of the other points are derived from these, for example northeast and south-southwest. Modern compasses are graduated into 360° rather than into cardinal points.

13   A student uses the following procedure to solve the triangle below.

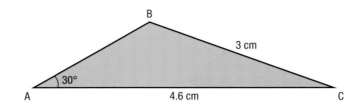

1. $\dfrac{3}{\sin 30°} = \dfrac{4.6}{\sin B}$, then m∠B = arc sin $\left(\dfrac{4.6 \times \sin 30°}{3}\right)$, which is ≈ 50°

2. m∠C = 180° − 30° − 50° = 100°

3. $\dfrac{3}{\sin 30°} = \dfrac{m\overline{AB}}{\sin 100°}$, then m $\overline{AB}$ = $\dfrac{3 \times \sin 100°}{\sin 30°}$, which is ≈ 5.91 cm

a)   Do the results obtained by the student seem right? Explain your answer.

b)   Identify and explain the error in this procedure made by the student.

c)   Solve the triangle above by correcting the error.

14   The bridge shown below enables pedestrians to cross a river. Using the information provided, determine the length of the bridge.

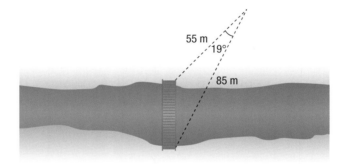

15   FORT JAMES In 1607, a triangular fort was built in Virginia; the purpose of the fort was to both ward off an attack by the Spaniards and to protect from the Powhatan Indians.

Archaeologists have recently discovered the remnants of this fort, part of which is immersed in the James River. The adjacent diagram represents the top view of the remaining part of Fort James.

a)   Determine the length of each of the original walls of the fort.

b)   Determine the original area of the fort.

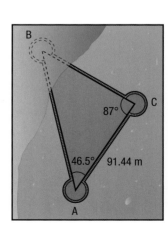

16 As shown below, the 4 bases of a baseball field correspond to the vertices of a square.

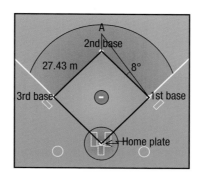

Considering that point A, 2nd base and home plate are perfectly aligned, determine the distance separating the first base from a player situated at point A.

17 The urban development plan depicted below shows the measurements obtained by two surveyors who need to determine a location for two new houses.

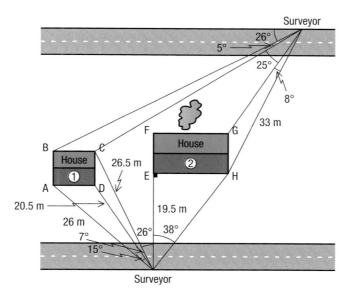

Determine:

a) the distance separating these 2 houses

b) the width of House ①, corresponding to m $\overline{AD}$

c) the depth of House ①, corresponding to m $\overline{CD}$

d) the width of House ②, corresponding to m $\overline{EH}$

e) the depth of House ②, corresponding to m $\overline{GH}$

Chronicle of the past

Hipparchus of Nicaea

His life

Considered to be the greatest astronomer of his time, Hipparchus is believed to have been born in 190 BCE, in Nicaea, Turkey. Based on the writings of many scholars, Hipparchus is said to have been the first to quantify and model the movements of the Sun and the Moon. Using an instrument of his own design called the astrolabe, he compiled a catalogue listing the positions of the stars. In addition to developing a method to predict solar eclipses, Hipparchus discovered the precession of equinoxes known as the conical wobbling of the Earth's axis of rotation once every 26 000 years. Hipparchus is also famous for his work in geometry.

He is credited with the creation of the first trigonometric tables. Historians believe Hipparchus died in 120 BCE in a city in ancient Greece.

The astrolabe is an instrument used to measure the angle of elevation of a celestial body in relation to a point of reference.

From the Earth to the Moon

With the help of trigonometry and astronomical observations during a total eclipse of the Moon, Hipparchus accomplished the extraordinary feat of calculating the distance between the Earth and its satellite by simply timing how long it takes for the Moon to cross the shadow of the Earth, and compared this time with the period of revolution of the Moon. Hipparchus also needed to know the diameter of the Earth, which one of his predecessors, Eratosthenes, had already calculated.

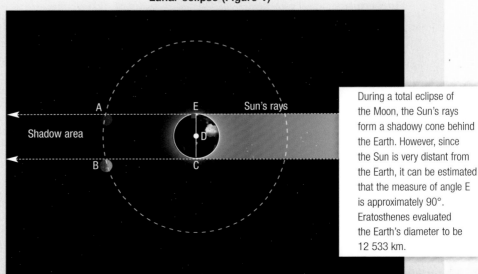

Lunar eclipse (Figure 1)

During a total eclipse of the Moon, the Sun's rays form a shadowy cone behind the Earth. However, since the Sun is very distant from the Earth, it can be estimated that the measure of angle E is approximately 90°. Eratosthenes evaluated the Earth's diameter to be 12 533 km.

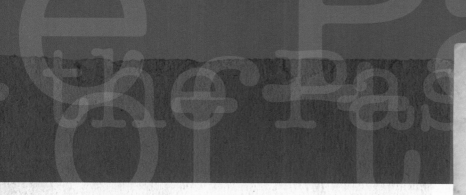

Table of lines inscribed in a circle

Arcs		Cords			Thirtieths of differences			
Degrees	Min.	Part of diam.	Prim.	Secon.	Part.	Prim.	Secon.	Tierc.
23	0	23	55	27	0	1	1	33
23	30	24	26	13	0	1	1	30
24	0	24	56	58	0	1	1	26
24	30	25	27	41	0	1	1	22
25	0	25	58	22	0	1	1	19
25	30	26	29	1	0	1	1	15
26	0	26	59	38	0	1	1	11
26	30	27	30	14	0	1	1	8
27	0	28	0	48	0	1	1	4
27	30	28	31	20	0	1	1	0
28	0	29	1	50	0	1	0	56
28	30	29	32	18	0	1	0	52
29	0	30	2	44	0	1	0	48
29	30	30	33	8	0	1	0	44
30	0	31	3	30	0	1	0	40
30	30	31	33	50	0	1	0	35
31	0	32	4	7	0	1	0	31
31	30	32	34	22	0	1	0	27

Tables and chords

Hipparchus is also known as the "father of trigonometry." His tables determined the length of a chord in a circle based on the angle subtended at the centre of the chord. The triangle formed can be divided into two congruent right triangles. Hipparchus' tables were used by Arab mathematicians and astronomers to create the trigonometric tables that we use today.

Circle (Figure 2)

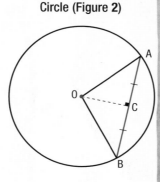

1. Considering that the total lunar eclipse (Figure **1**) observed by Hipparchus lasted 196.2 min, and that the moon's period of revolution is 27 days, 7 h and 43 min, calculate:

a) the measure of angle ADB

b) the measure of angle ADE

c) according to Hipparchus' method, the distance from the Earth to the Moon is represented by segment AD

2. Presuming that the circle in Figure **2** has a radius of 1 dm, complete the following table without using a calculator.

Measure of angle AOB subtended at the centre (°)	Length of chord AB (dm)	Measure of angle AOC subtended at the centre (°)	Sine of angle AOC $\frac{m \overline{AC}}{m \overline{AO}}$
0	0.00	0	0
10	0.174	5	0.087
20	0.347	10	0.1735
30	0.518	15	

3. Based on the results obtained in **2.**, explain how ancient Arab mathematicians were able to construct a sine table, based on Hipparchus' chord tables.

In the workplace

Navigation officers

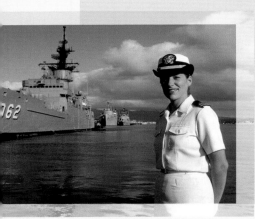

The job

A navigation officer's job consists of operating a variety of ships carrying cargo or passengers. These officers might work for a transport company, the tourist industry, the Canadian Coast Guard or the public service.

Tasks and responsibilities

The tasks related to navigation are extremely diverse and depend on the officer's rank. Along with preparing cargo plans, checking weather forecasts and making sure that all onboard instruments work properly, a navigation officer must be able to read nautical charts and ensure the safety of the crew and cargo. With time and experience, he or she can attain the rank of master mariner or commander.

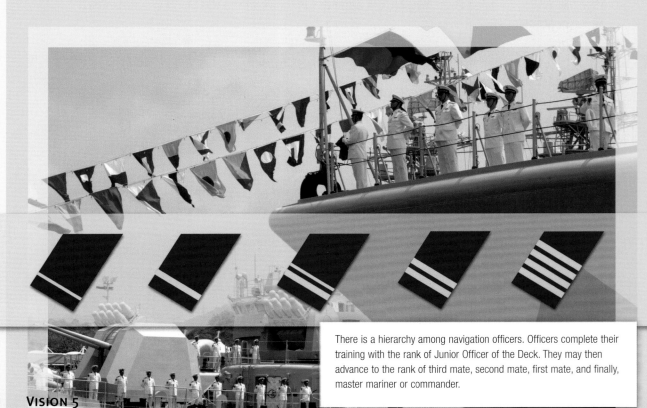

There is a hierarchy among navigation officers. Officers complete their training with the rank of Junior Officer of the Deck. They may then advance to the rank of third mate, second mate, first mate, and finally, master mariner or commander.

Where are we?

Navigation officers must also be able to determine a ship's position at any given moment. They must be familiar with electronic positioning and traditional methods in case of electronic breakdown. In both cases, these methods involve the concepts of trigonometry.

Navigation techniques differ according to whether one is in sight of land or not. Celestial navigation is based on star sightings; coastal navigation requires the observation of landmarks which are fixed points along the shore with known positions on the charts. Navigation by dead reckoning allows you to deduce the ship's position based on its speed, its bearing and its last known position. In the illustration below, you can see two landmarks that a ship might use to determine its position.

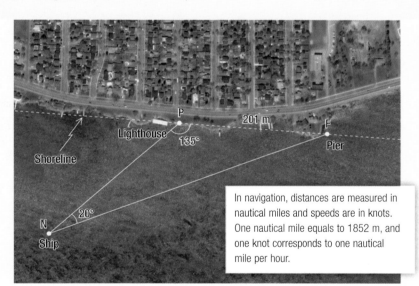

In navigation, distances are measured in nautical miles and speeds are in knots. One nautical mile equals to 1852 m, and one knot corresponds to one nautical mile per hour.

Not just at sea

Navigation officers' responsibilities are not confined to operating a ship. They may also be called upon to supervise and plan cargo loading and unloading operations. Among other things, officers must optimize the use of all available space on the ship and take into account how the placement of the load will affect navigation.

1. Answer the following questions using the information provided in the adjacent illustration:

a) How far is the ship from the lighthouse?

b) How far is the ship from the pier?

c) Considering that the ship cannot be closer than 185 m from the shore, should the officer move the ship away from the shore? Justify your answer.

2. A ship leaves its home port at point A to move in a straight line towards a destination point B that is 230 nautical miles away. After navigating for 2.5 h at a speed of 20 knots, the officer realizes that, from the start of the trip, a northwesterly wind has caused the ship to deviate 7° off course.

a) What angle is required to correct the course of the ship to allow it to sail to its final destination in a straight line?

b) How many kilometres will the ship deviate from its initial course?

c) Verify the answers obtained in **a)** and **b)** by representing the situation in a scale diagram (1 cm ≙ 20 nautical miles).

1 For each of the following triangles, solve for *x* and *y*.

a)
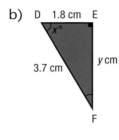

b)
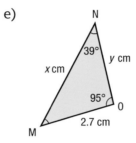

c)
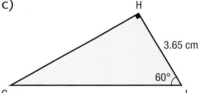

d)
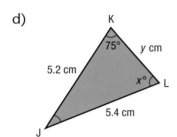

e)
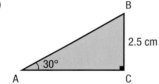

f)
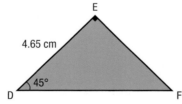

2 Determine the area of each of the triangles below.

a)

b)

c)

d)
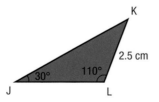

e)
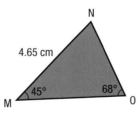

f)
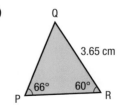

3 In a port, employees are busy loading cars onto a ship. If the loading ramp has an inclination of 15° and its upper end is at a height of 5 m above the dock, determine the length of this ramp.

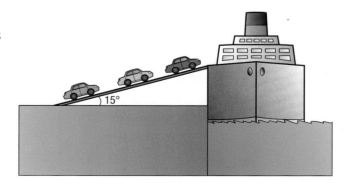

4 Calculate the area of the triangle whose lengths are:

a) 3.4 cm, 5.2 cm, 7.1 cm

b) 57 mm, 102 mm, 140 mm

c) 0.8 km, 1.1 km, 1.5 km

d) 123.4 cm, 3456 mm, 0.42 dam

5 To give mobility-impaired people access to a building, a ramp is constructed in the shape of a triangular-based right prism that is 1.5 m wide. This ramp is built next to a stairway that is 1.2 m high. Building standards require that the access ramp have a 1:16 maximum ratio of height to depth.

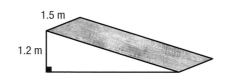

a) What is the angle of elevation of the ramp?

b) What is the volume of this structure?

c) The contact surface of the ramp is finished in anti-skid paint. Calculate the area of the surface to be covered.

6 A 20-m tall lighthouse is topped with a light that revolves at an angle varying from 0° to 50° from a vertical position. Not counting the actual structure of the lighthouse:

a) what is the area of the illuminated surface on the ground?

b) how much space is illuminated by the light?

7 Three markers are used to determine the horizontal distance between two points located on either side of a hill. These markers make it possible to determine the lengths shown on the diagram below.

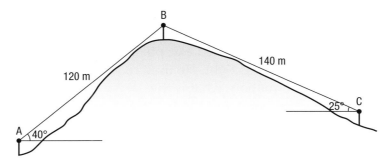

a) What is the horizontal distance between Markers **A** and **C**?

b) What is the vertical distance between Markers **A** and **C**?

c) What is the distance between Markers **A** and **C**?

8 What is the volume of the adjacent regular pyramid?

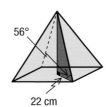

9 A water-table detection device located at point A indicates that there is a spring just underneath the device. Drillers realize that it is impossible to reach the spring by drilling at point A, so they decide to drill from point B in the direcon of the spring.

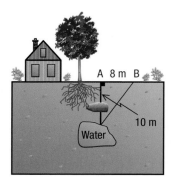

a) At what distance from the spring is point B located?

b) Using a vertical line as a basis, at what angle must the drillers drill in order to reach the spring?

10 During a hot air balloon show, a woman gazes eastward, at an angle of elevation of 52°. She is looking at a hot air balloon that is moving at a constant speed in a horizontal direction. Three minutes later, the woman looks at the same hot air balloon, but her gaze is westward at an angle of elevation of 45°. If the hot air balloon is at an altitude of 1500 m, at what speed is it moving?

In 2008, the International Balloon Festival of St-Jean-sur-Richelieu celebrated its 25th anniversary. This festival is the biggest gathering of balloons in Canada with more than a hundred hot air balloons and an average of 350 000 visitors each year.

11 An electrician must install an underground cable so that it passes through a small hill. To complete this task, he attaches a rope at the intended entry point, attaches the same rope at the intended exit point and then pulls the rope until it is tight. He uses 13 m of rope. Based on this information, what is the distance between the entry point and the exit point of the cable?

Rope

50°   34°

Cable entry   Cable exit

12 According to a Chinese legend, General Han Xi wanted to invade the royal palace and take over as leader. To fulfill his plan, he used a kite to measure the distance between his home and the palace, and then he dug a tunnel to gain entry.

If he used 300 m of tether line to fly his kite at a 25° angle of elevation, what was the length of his tunnel?

The Chinese used kites for military purposes such as sending signals, asking for reinforcements and scaring the enemy by flying kites with bamboo whistles.

13 **CABLE CAR** In France, there is a cable car from Chamonix to the top of Aiguille du Midi, on Mont Blanc. The table below contains some information on this topic:

Aiguille du Midi cable car

	1st stage	2nd stage
Length (m)	2553	2867
Vertical difference (m)	1279	1470
Base station altitude (m)	1038	
Top station altitude (m)		

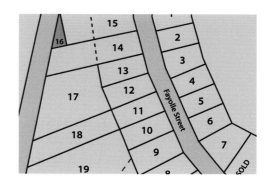

Aiguille du midi

Relay station

Chamonix

a) Complete the table above.

b) What is the angle of elevation from the horizon:

1) through the 1st stage?

2) through the 2nd stage?

c) A cable car cable is extended directly from Chamonix to Aiguille du Midi. What is the length of this cable?

14 The adjacent plan shows the lots for sale in a residential development. The sides of the only triangular lot available measure 30.1 m on the east side, 10.67 m on the south side and 32.92 m on the northwest side. How much does this lot cost if each square metre sells for $1,250?

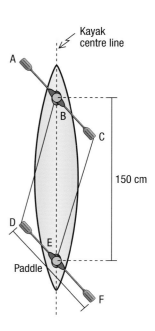

15 The paddles of two kayakers are represented by segments AC and DF in the adjacent diagram. They each measure 200 cm. For safety reasons, the distance separating one kayaker from the end of the other kayaker's paddle cannot be less than 75 cm.

a) What is the maximum measure of angle DEB at which the kayaker can hold his paddle while complying with these restrictions?

b) If this same kayaker holds his paddle so that angle DEB is 45°, what is the length of the longest paddle he can use?

Kayak centre line

A

B

C

150 cm

D

E

Paddle

F

16 **THE FLATIRON BUILDING** The Flatiron Building, located in New York City, is in the shape of a right prism with a base in the shape of a right triangle with rounded vertices. The diagram below shows the top view of this building.

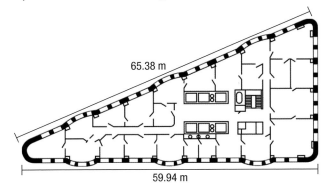

The unusual shape of this building causes unexpected wind currents on New York's 23rd Street.

a) What is the length of the shortest side of the building?

b) What is the measure of each angle defined by the sides of the building?

17 The construction of the Egyptian Pyramids required many steps. As shown in the illustration below, cut stones had to be carried from the quarry to the pyramid construction site.

a) During Step **1**, the Egyptians had to transport the stones 25 m at an angle of inclination of 22° from the horizon. At what height were these stones before they were moved down the inclined path?

b) During Step **5**, if the Egyptians had to walk 40 cm/sec for 25 min to transport the stones up the inclined path. Determine:

1) the distance covered

2) the angle of elevation of the path if the top of the path was at a height of 8 m

18 As shown in the adjacent illustration, a tree projects a 15-m shadow on the side of a mountain. Considering that the side of this mountain has an inclination of 18° from horizon, calculate the height of this tree.

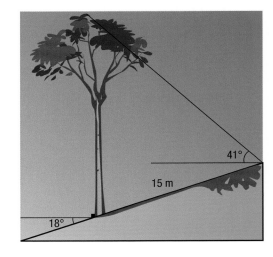

19 To map a given location, surveyors need a reference distance as well as a device to measure angles. They divide the surface to be surveyed into triangles and place markers at their vertices. Considering that the distance between Markers **E** and **D** is 75 m:

a) determine the distance between Markers **B** and **F**

b) calculate the total area of the terrain

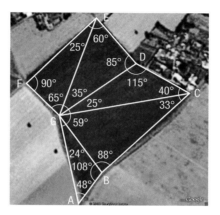

20 When light passes from one transparent environment to another, it undergoes a certain amount of angular deflection which produces virtual images. These images may seem distorted or displaced in reference to reality as shown in the image below.

a) From the diver's perspective, at what height above the water is the top of the observer's head?

b) From the observer's perspective, the diver is viewed at a distance *d* from his actual position. What is this distance?

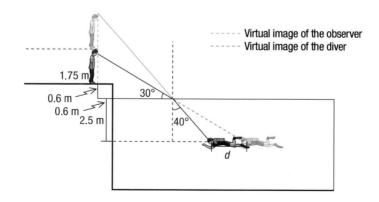

The deflection of light as it changes from one environment to another is called refraction. In the adjacent picture, the pen seems cut in two. This is an optical illusion created by the formation of a virtual image in the water due to light refraction.

21 ANGLE OF CONVERGENCE When looking at an object that is relatively close, your eyes converge in such a way that your visual axes align with the object. Your brain evaluates the distance of the object based on the convergence. The illustrations below represent the eyes of two people that are observing the same object located the same distance away.

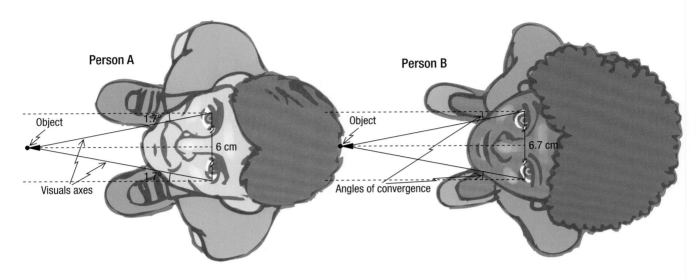

a) What is the distance between the object and Person **A** and **B**?

b) What is the angle of convergence of the eyes of Person **B**?

c) A person observes an object 95 cm in front of their eyes. To do this, their eyes each converge 1.45°. Calculate the distance between this person's eyes.

Presbyopia is a common condition for people aged 45 and up. It is characterized by difficulty reading at close range. With age, the crystalline lens, a structure that allows the eye to focus, loses some of its flexibility, which makes close-range vision more difficult.

22 EARTH-MOON The diagram below shows the arrangement of the Earth-Moon system when the latter is the least distant from the Earth. Based on the information provided, determine the diameter of the Moon.

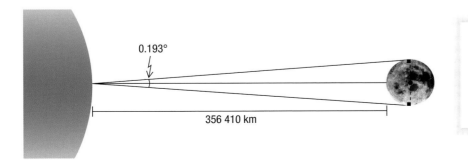

0.193°

356 410 km

The Moon follows an elliptical trajectory around the Earth. When the Moon is situated at its point closest to the Earth, it is at its perigee. At its most distant point, it is at its apogee.

23 THE EQUESTRIAN Sculpted by Croatian artist Antun Augustincic (1900-1979), this statue was offered as a gift to the United Nations by the government of Yugoslavia in 1954. When it was restored in 2008, the project managers referred to the diagram below in order to determine the height of the statue. This diagram shows the angle of incidence of the Sun's rays according to date and time of day.

Considering that, at 2:00 p.m. on August 21, 2008, the shadow of the statue measured 367.49 cm, determine the height of the statue.

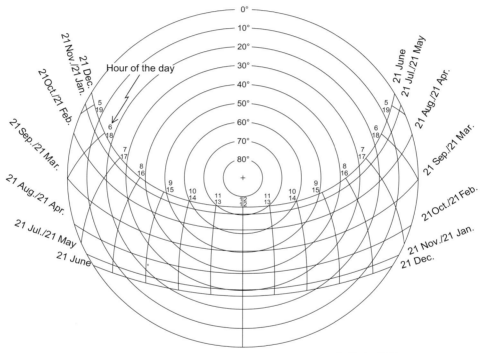

Latitude 40 N

Rosanne Dubé is a private architect who is part of the team with the mandate to relocate and restore the statue called "The Equestrian." A specialist in heritage conservation, she had to determine the appropriate methods to move the statue and ensure that both the statue and its base could be reset according to its original 1954 dimensions.

bank of problems

24 The information below was collected by an individual who observed a grain silo in the shape of a right circular cylinder.

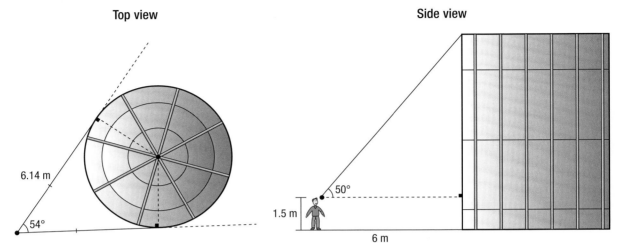

Top view

Side view

6.14 m

54°

50°

1.5 m

6 m

Show that the volume of this silo is approximately 266.23 m³.

25 Pedro wants to measure the distance between the extremities A and B of a lake. He proceeds in the following manner. From point O, he takes note of various angles. He then moves based on a known angle until he reaches point P located a certain distance away from point O. From point P, he starts measuring other angles.

The adjacent diagram indicates all the measurements that Pedro found. Based on the information provided, determine the length of the lake.

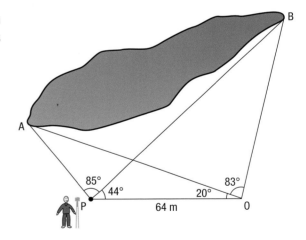

B

A

85° 44°   20° 83°

P   64 m   O

26 A sailboat skipper needs to order a sail identical to the one in the adjacent illustration. Give the skipper the necessary instructions to calculate the dimensions and the area of the sail that needs to be ordered if the only known measurement are those of the mast AC, the boom BC and angle DBC.

A

027

D

C   B

27 Geodesy is the science of studying the shape and dimensions of the Earth. In geodesy, trigonometric levelling is often used to determine the altitude of geodesic benchmarks.

In the diagram below, considering that point A is located 1200 m above sea level, determine the altitude of point B.

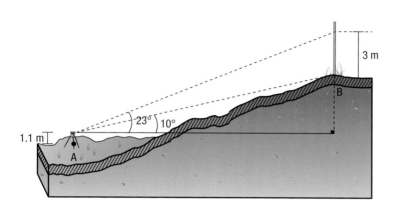

A geodetic marker allows surveyors to know the exact coordinates of the marker and use it to make measurements of other nearby objects or locations.

28 Two parachutists jump from an airplane flying at 250 km/h, at an altitude of 4000 m directly above point O, which is the expected landing spot. Because of wind gusts of approximately 45 km/h, the first jumper experiences a vertical deviation of 15° and the second experiences a vertical deviation of 22°. Once on the ground, Parachutist **A** must join up with Parachutist **B**.

Determine the angle, in reference to the north, that Parachutist **A** should take as well as the distance she will travel.

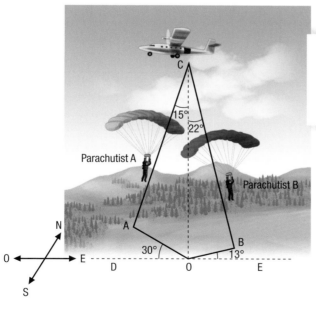

The direction individuals must take during an orientation exercise is called the azimuth. The azimuth is the angular distance from the North that individuals must apply to their trajectory.

VISI⑥N

Probability of random experiments

Is it possible to predict or even control chance? How is the probability of rain forecasted? How do you know when a game of chance is fair or not? To what extent does probability influence the expectations of financial gain for certain companies? In "Vision 6," you will determine various probabilities by playing the role of a meteorologist and financial analyst. With the help of mathematical expectation, you will determine whether a game of chance is fair. If the game is unfair, you will make modifications in order to make it fair.

Arithmetic and algebra   Geometry   Statistics   **Probability**

- Enumerating possibilities
- Subjective probability
- Odds for and odds against
- Distinction between odds and probability
- Mathematical expectation and fairness

PRIOR LEARNING 1 A litter of kittens

A cat is about to give birth to two kittens. Based on the features of the kittens' parents, the veterinarian estimates a 25% probability of a kitten being born with white fur and a 75% probability of another kitten being born with spotted fur. The probability of the cat giving birth to a male kitten is estimated at 50%.

A female cat can reproduce as of the age of six months. The gestation period lasts approximately two months, and on average, a litter consists of four kittens.

a. Complete the following probability tree.

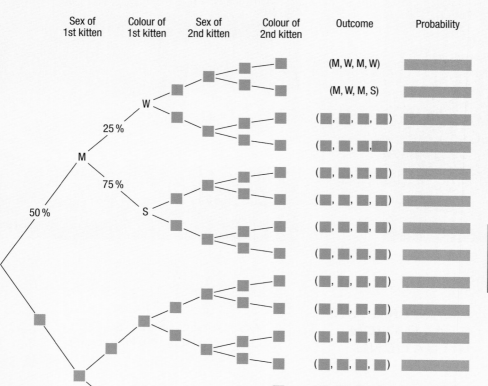

Sex of 1st kitten	Colour of 1st kitten	Sex of 2nd kitten	Colour of 2nd kitten	Outcome	Probability

Legend
M : Male
F : Female
W : White
S : Spotted

b. Determine the probability that:

1) both kittens will have spotted fur

2) both kittens will be male

3) one of the two kittens will be male and have white fur

At an outdoor recreational centre, members can sign up for any combination of three activities: tree adventures, canoeing and hiking. The following is a sign-up list of the 46 members participating in different activities in the course of one day:

Sign-up list	
Activity:	Number of people:
Tree adventures	20
Canoeing	16
Hiking	25
Canoeing and tree adventures	5
Canoeing and hiking	9
Tree adventures and hiking	10
Tree adventures, hiking and canoeing	3
No activity	6

a. How many people are signed up for only:

1) tree adventures?

2) canoeing?

3) hiking?

b. One person is chosen at random from the outdoor recreational centre that day. What is the probability of choosing a person signed up for:

1) canoeing?

2) more than one activity?

3) tree adventures or canoeing?

4) tree adventures and canoeing?

George Hébert, a Frenchman, invented a method of physical training based on a series of outdoor exercises. These exercises are done on special obstacle courses made of wood and rope.

knowledge summary

RANDOM EXPERIMENT

An experiment is **random** if the following is true:

1. Its outcome depends on **chance**, meaning that the outcome of the experiment cannot be predicted with absolute certainty.

2. Before the experiment, the set of all the possible outcomes which is called the **sample space** can be listed; this set is indicated as "Ω" otherwise known as "omega."

> E.g. A marble is drawn from a bag containing 10 marbles of the same size; each one is numbered from 1 to 10. The sample space is Ω = {1, 2, 3, 4, 5, 6, 7, 8, 9, 10}.

EVENT

An **event** is a **subset** of sample space. An event is considered **simple** if it consists only of one **outcome** from the sample space.

> E.g. 1) When drawing a card from a deck of 52 cards, "drawing an ace" is an event and corresponds to {ace of hearts, ace of spades, ace of diamonds, ace of clubs}.
>
> 2) When tossing a coin, "landing on heads" is a simple event because it represents only one outcome {heads} of the sample space.

COMPATIBLE AND INCOMPATIBLE EVENTS

Events A and B are **compatible** if they can occur at the same time, in other words if $A \cap B \neq \emptyset$.

Events A and B are **incompatible** if they cannot occur at the same time, in other words if $A \cap B = \emptyset$.

> E.g. Rolling a six-sided die. The events "roll an even number" and "roll a number greater than 3" are compatible events.

> E.g. Rolling a six-sided die. The events "roll a multiple of 5" and "roll a number less than 4" are incompatible events.

COMPLEMENTARY EVENTS

Events A and B are **complementary** if they are incompatible and if the combination of possible outcomes of both events represents the entire sample space.

If $A \cap B = \emptyset$ and $A \cup B = \Omega$, events A and B are complementary.

If A and B are complementary, then $P(A) + P(B) = 1$.

> E.g. Roll a six-sided die, numbered from 1 to 6 and observe the side facing up. The events "rolling an even number" and "rolling an odd number" are complementary.

PROBABILITY OF AN EVENT

The **probability of an event** composed of several simple events is equal to the **sum of the probabilities** of each simple event.

E.g. A drawer contains 8 knives, 10 forks and 12 spoons. Since "taking out a knife at random" and "taking out a fork at random" are two simple events, the probability of the event "taking out a knife or a fork at random" is written as follows:

$$P(\text{knife or fork}) = P(\text{knife}) + P(\text{fork}) = \frac{8}{30} + \frac{10}{30} = \frac{18}{30} = \frac{3}{5}$$

RANDOM EXPERIMENT WITH SEVERAL STEPS

In a random experiment with several steps, the probability of an event is equal to the **product of the probabilities** of each intermediate event at each step that forms this event.

E.g. Two marbles are drawn from a bag containing 10 marbles that are the same size and numbered from 1 to 10. The probability of the event "draw marble 3 followed by marble 7" is written as follows:

- if the marble is put back into the bag after the first draw, which is with replacement,

$$P(3 \text{ followed by } 7) = P(3) \times P(7) = \frac{1}{10} \times \frac{1}{10} = \frac{1}{100}$$

The first marble was put back in the bag.

- if the marble is not put back into the bag after the first draw, which is without replacement,

$$P(3 \text{ followed by } 7) = P(3) \times P(7) = \frac{1}{10} \times \frac{1}{9} = \frac{1}{90}$$

The first marble was not put back in the bag.

The following tree diagram illustrates all the possible outcomes of an experiment with several steps. By including a probability to each branch in the tree diagram, you get a **probability tree diagram**.

E.g. You toss a coin twice.

Number of possible outcomes:
$2 \times 2 = 4$

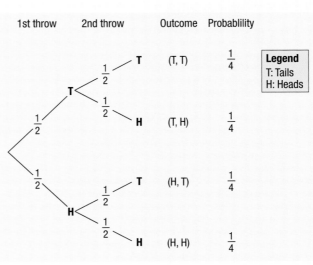

1st throw	2nd throw	Outcome	Probability
	T	(T, T)	$\frac{1}{4}$
T	H	(T, H)	$\frac{1}{4}$
H	T	(H, T)	$\frac{1}{4}$
	H	(H, H)	$\frac{1}{4}$

Legend
T: Tails
H: Heads

knowledge in action

1. List the sample space for each of the following situations.

   a) Toss a coin and take note of the side facing up.

   b) Roll a six-sided die numbered from 1 to 6 and take note of the side facing up.

   c) Pick a ball from a set of red, green, blue and white balls and take note of its colour.

   d) Draw a card from a deck of 52 cards and take note of the colour of the card.

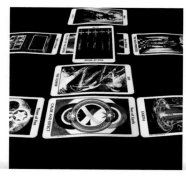

There are numerous types of playing cards. In fact, there are seven museums in the world that are devoted to cards. Many cards were invented specifically for certain board games.

2. A bag contains 4 red marbles, 2 green marbles, 6 white marbles and 3 blue marbles. What is the probability of randomly picking:

   a) a red marble?   b) a green marble?   c) a red or green marble?

   d) a green marble followed by a red marble if you put the marble back in the bag after the first draw?

   e) a green marble followed by a red marble if you do not put the marble back in the bag after the first draw?

3. Pool is played on a felt table with 16 balls: 1 unnumbered white ball and 15 numbered balls. Balls 1 to 8 are solid colours and balls 9 to 15 have a coloured stripe.

   If you choose a ball at random from these 16 balls, what is the probability of choosing:

   a) an even-numbered ball?

   b) a blue ball?

   c) a striped ball or the eight ball?

   d) a red or black ball?

   e) a yellow, purple or brown ball?

Ball number	Colour
1 and 9	Yellow
2 and 10	Blue
3 and 11	Red
4 and 12	Purple
5 and 13	Orange
6 and 14	Green
7 and 15	Brown
8	Black

There are many versions of American billards or pool. In the most commonly played version, each player chooses either solid coloured balls or striped balls. The player who first sinks all of his or her balls and then sinks the black ball wins the game.

4 Jasmin uses a hole punch to make confetti by punching 50 holes in a pile of 4 white sheets, 5 pink sheets, 3 blue sheets and 4 red sheets. She consecutively picks, at random and without replacement, one piece of confetti and then another.

a) Draw a probability tree associated with the colour of the confetti.

b) Calculate the probability of picking:

   1) two red pieces of confetti

   2) two pieces of confetti of the same colour

   3) two pieces of confetti of different colours

In Italian, *confetti* means "sugar-coated almond." Originally, party-goers would throw candies which then were replaced by little bits of plaster. Today the tradition is to throw little specks of paper.

5 A bag contains 10 red marbles numbered from 0 to 9, 10 white marbles numbered from 0 to 9 and 10 blue marbles numbered from 0 to 9. You randomly choose 4 marbles consecutively with replacement. Calculate the probability that:

a) all the chosen marbles are of the same colour

b) all the chosen marbles are red

c) the first chosen marble is marked "1," the following three are marked "0"

d) the first chosen marble is marked "1," the following three are marked "0" and they are all of the same colour

e) the first chosen marble is marked "1," the following three are marked "0" and they are all red

6 After a soccer game, the coach hands out frozen treats. Leo is second in line. Initially, the cooler held 6 treats of each of the following flavors: grape, cherry, orange and lime. If the coach hands out the frozen treats at random, what is the probability of Leo getting a grape treat?

7 FRS/GMRS portable radios operate on 22 frequency channels and can use 38 different access codes. In order for two people to communicate with each other, they must use the same frequency channel and access code. If Tina tunes her radio to channel 3 and chooses access code 4, what is the probability that another person will:

a) randomly tune in to the same frequency channel?

b) randomly choose the same access code?

c) randomly tune in to the same frequency channel and randomly choose the same access code?

FRS stands for *Family Radio Service* and GMRS stands for *General Mobile Radio Service*. These low-powered, short-range devices are very practical for people who enjoy outdoor activities.

SECTION 6.1 Enumerating possibilities

This section is related to LES 12.

PROBLEM Scrabble

Scrabble was invented by New York architect, Alfred Mosher Butts. The original creation, called "Lexico," was released in 1931. Unlike Scrabble, Lexico was not played on a board, but people nevertheless kept score when forming words. Alfred Butts analyzed the front page of the *New York Times* in order to calculate the frequency of each letter as a means of assigning each one a point value. In the current English version of Scrabble, there are 100 tiles as shown on the right. The blank tile can substitute for any letter.

According to the official rules, after determining who goes first, each person picks seven letters and then the game begins.

Letter	Frequency	Value
A	9	1
B	2	3
C	2	3
D	4	2
E	12	1
F	2	4
G	3	2
H	2	4
I	9	1
J	1	8
K	1	5
L	4	1
M	2	3
N	6	1
O	8	1
P	2	3
Q	1	10
R	6	1
S	4	1
T	6	1
U	4	1
V	2	4
W	2	4
X	1	8
Y	2	4
Z	1	10
Blank	2	0

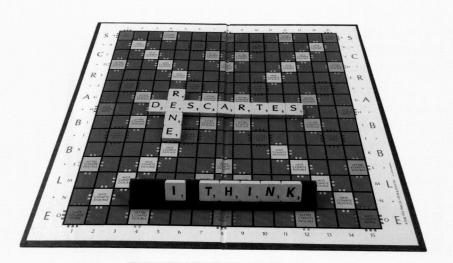

Scrabble is sold in over 120 countries in 29 different languages.

If the first player that picks tiles can arrange them without including the blank one, what is the probability that the word "JUPITER" will be spelled during the first round?

ACTIVITY 1 A magic trick

During a magic show, a magician asks a member of the audiance to shuffle a deck of 52 cards and place them in a hat.

a. In a deck of 52 cards, how many are:

1) aces?

2) hearts?

3) ace of hearts?

A second member of the audiance blindfolds the magician who then consecutively draws 13 cards from the hat and places them on the table. The adjacent illustration shows the arrangement of the cards that he drew. Suppose this arrangement of cards was chosen at random.

b. Fill in the following table.

Draw	1	2	3	4	5	6	7	8	9	10	11	12	13
Number of cards remaining in the deck													

c. While the magician consecutively draws the 13 cards, what is the probability that he draws:

1) the ace of hearts first?

2) the two of hearts second?

3) the three of hearts third?

d. Keeping in mind the order of the cards, how many sets of 13 cards can the magician draw?

e. What is the probability of drawing, in the same order, the arrangement shown above if you select at random and without replacement, 13 cards in a deck of 52 cards?

Houdini has always been known as one of the most famous magicians. He was born in Hungary in 1874, and his family immigrated to the United States when he was four years old. Around the age of 17, he began performing at fairs as a magician before establishing himself as a master escape artist capable of freeing himself from any restraint. He died in 1926.

During a game of Bingo, each player receives a different card that has a grid of 5 by 6 squares. The first row of squares contains the letters B-I-N-G-O and the other squares, except for from the centre square, each has a different number ranging from 1 to 75. Simon receives a card such as the one shown in the adjacent illustration.

B	I	N	G	O
11	20	31	50	61
1	18	40	46	70
5	25	Free space	57	74
12	16	45	59	68
14	30	35	48	63

The Bingo caller picks the numbered balls consecutively, without replacement, in the following way:

B	I	N	G	O
1 to 15	16 to 30	31 to 45	46 to 60	61 to 75

a. What is the probability that ball B-11 is:
   1) the first ball selected?
   2) the second ball selected?

b. In how many ways is it possible that the first 5 balls selected are the balls B-11, I-20, N-31, G-50, O-61?

c. What is the probability that the first five selected balls are B-11, I-20, N-31, G-50 and O-61:
   1) in this order?
   2) in any order?

d. 1) What is the probability of Simon completing the row shown on the right after only 5 balls are selected?

   2) Does the order in which the balls are selected have an effect on this probability? Explain your answer.

B	I	N	G	O
1	18	40	46	70
5	25	Free space	57	74
12	16	45	59	68
14	30	35	48	63

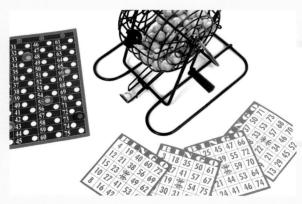

Bingo is a game of chance that dates back to the 1530s in Italy. The ancient version of the game spread all the way to France, then to Germany in the 19th century where it was used in schools to help children learn the alphabet. It was only at the start of the 20th century in the United States that Bingo finally became the game we know today.

Techno math

Many calculators allow you to quickly solve enumeration problems using permutations, combinations and factorials of numbers.

This screen shows the various functions available when calculating probabilities.

Screen 1

```
MATH NUM CPX PRB
1:rand
2:nPr
3:nCr
4:!
5:randInt(
6:randNorm(
7:randBin(
```

This screen shows three permutation calculations.

Screen 2

```
4 nPr 2
            12
9 nPr 7
        181440
2 nPr 2
             2
```

This screen shows three combination calculations.

Screen 3

```
4 nCr 2
             6
9 nCr 7
            36
2 nCr 2
             1
```

This screen shows the factorial of three numbers.

Screen 4

```
4!
            24
5!
           120
8!
         40320
```

a. After comparing the adjacent screen with Screen **4**, explain what the factorial function allows you to calculate.

b. Referring to Screen **2**:
1) how many ways are there to choose 2 elements from 4 if order is taken into account?
2) what do the two numbers on either side of nPr represent?

c. Referring to Screen **3**:
1) how many ways are there to choose 7 elements from 9 if order is not taken into account?
2) what do the two numbers on either side of nCr represent?

d. Why are the last two results from Screens **2** and **3** different?

e. Are the permutations and combinations calculated by the calculator with or without replacement? Justify your answer.

f. Using a calculator, determine:
1) the number of possible teams of 11 people that can be created by a group of 30
2) the number of 9-character access codes that can be created without repetition

knowledge 6.1

FACTORIAL

The study of arrangements, permutations and combinations often use the **factorial** operation. The factorial of a natural number n is written as $n!$ and is calculated as:

$$n! = n \times (n-1) \times (n-2) \times (n-3) \times \ldots \times 3 \times 2 \times 1$$

E.g. The factorial of $8 = 8! = 8 \times 7 \times 6 \times 5 \times 4 \times 3 \times 2 \times 1 = 40\ 320$.

Keep in mind that $0! = 1$.

RANDOM EXPERIMENTS WITH OR WITHOUT ORDER

In a **random experiment**, you can either take into account the order of the outcome or not. When you do not take the order of the outcome into account, the sample space generally tends to have fewer results.

PERMUTATION AND ARRANGEMENTS OF A WHOLE SET

The **permutation** of a set of elements corresponds to an **ordered sequence** of elements of this set. Two permutations are distinguished by the order in which the elements of this set are arranged.

E.g. The set of letters {A, B, C} can be written in 6 different ways: (ABC), (ACB), (BAC), (BCA), (CAB) and (CBA).

The number of permutations of a set of n elements can be determined as follows:

Number of possible elements for the 1st position	×	Number of possible elements for the 2nd position	×	Number of possible elements for the 3rd position	× ... ×	Number of possible elements for the nth position	$= n!$

E.g. The number of possible permutations for the set of letters {A, B, C, D} is:

Number of possible elements for the 1st position	×	Number of possible elements for the 2nd position	×	Number of possible elements for the 3rd position	×	Number of possible elements for the 4th position	$= 4 \times 3 \times 2 \times 1 = 4! = 24$

ARRANGEMENT

An **arrangement**, which is also a permutation, of a set of elements corresponds to the **ordered position of a certain number of elements** of this set. Two arrangements are distinguished by the order in which the elements are arranged.

To determine the number of arrangements of a set containing r elements chosen from a set containing n elements:

Number of possible elements for the 1st position	×	Number of possible elements for the 2nd position	×	Number of possible elements for the 3rd position	× ... ×	Number of possible elements for the rth position

E.g. Two numbers are chosen randomly from the set {1, 2, 3, 4, 5}.

If the random experiment is carried out without replacement, there are 5 possible elements for the 1st number and 4 possible elements for the 2nd number. Therefore, there are 5 x 4 = 20 possible arrangements:

(1, 2), (1, 3), (1, 4), (1, 5), (2, 1), (2, 3), (2, 4), (2, 5), (3, 1), (3, 2), (3, 4), (3, 5), (4, 1), (4, 2), (4, 3), (4, 5), (5, 1), (5, 2), (5, 3), (5, 4)

If the random experiment is carried out with replacement, there are 5 possible elements for the 1st number and 5 possible elements for the 2nd number. Therefore, there are 5 x 5 = 25 possible arrangements.

COMBINATION

A **combination** of a set of elements corresponds to an **unordered position** of a certain number of elements of this set.

You can determine the number of possible combinations of a random experiment without replacement as follows:

$$\text{Number of possible combinations} = \frac{\text{number of possible arrangements}}{\text{number of permutations}}$$

E.g. 1) Three-lettered words are formed at random without replacement using letters A, B, C and D. The number of possible combinations is determined as follows:

$$\text{Number of possible combinations} = \frac{\text{number of possible arrangements}}{\text{number of permutations}} = \frac{4 \times 3 \times 2}{3 \times 2 \times 1} = \frac{24}{6} = 4$$

Keeping in mind the order, there are 24 possible results:

(A, B, C), (A, B, D), (A, C, B), (A, C, D), (A, D, B), (A, D, C), (B, A, C), (B, A, D), (B, C, A), (B, C, D), (B, D, A), (B, D, C), (C, A, B), (C, A, D), (C, B, A), (C, B, D), (C, D, A), (C, D, B), (D, A, B), (D, A, C), (D, B, A), (D, B, C), (D, C, A), (D, C, B)

There are 6 different ways of writing a word formed by the same letters if you take into account the order:

(A, B, C), (A, C, B), (B, A, C), (B, C, A), (C, A, B), (C, B, A)

Therefore, the possible combinations are:

(A, B, C), (A, B, D), (A, C, D), (B, C, D)

2) Three balls are drawn at random and without replacement from a vase containing 5 balls: 1 red, 1 blue, 1 green, 1 orange and 1 yellow. The number of possible combinations is determined as follows:

$$\text{Number of possible combinations} = \frac{\text{number of possible arrangements}}{\text{number of permutations}} = \frac{5 \times 4 \times 3}{3!} = \frac{60}{6} = 10$$

practice 6.1

1 Calculate the results of the following operations.

a) 4!   b) 8!   c) 10!   d) 1! + 10!

e) 0!   f) $\frac{15!}{14!}$ g) 5! × 7!   h) $\frac{6! - 4!}{(6-4)!}$

Blaise Pascal invented the first calculator in 1640. This calculating machine, called Pascaline, was mechanical and could only add and subtract.

2 The factorial function generates large numbers very quickly. As shown in the adjacent screen, it can even exceed the capacity of certain calculators.

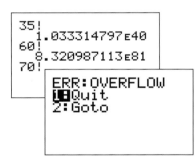

Calculate the results of the following operations:

a) $\frac{105!}{100!}$ b) $\frac{98!}{(98-5)! \times 5!}$

3 Determine how many different ways it is possible to write these sets.

a) {A, B, C, D, E}   b) {2, 4, 6}   c) {tails, heads}   d) {1, 2, 3, 4, 5, 6}

e) {hearts, spades, diamonds, clubs}   f) {ace, 2, 3, 4, 5, 6, 7, 8, 9, 10, jack, queen, king}

4 If you randomly choose 6 out of the 26 letters of the alphabet, how many different six-letter "words" can you write even if they don't make sense:

a) if a letter can only be chosen once?

b) if a letter can be chosen more than once?

5 How many different ways can a group of 12 people line up in a single file?

6 If you consider the two situations shown in the adjacent illustrations as identical, how many different ways can a group of 5 people sit around a round table?

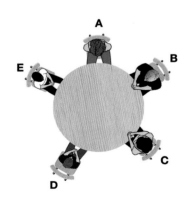
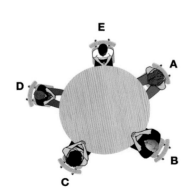

7 The following children's puzzle is made up of little pieces that once correctly assembled form a picture. How many different ways can the pieces of this puzzle be placed?

8 A factory produces purple, blue, yellow and red candies. As shown in the adjacent illustration, four candies are placed in a tube creating an assortment.

a) How many different assortments can be created:
1) if one candy of each colour is placed in the tube?
2) if more than one candy of each colour is placed in the tube?

b) If more than one candy of each colour can be placed in the tube and one assortment is randomly chosen out of all the possible assortments, what is the probability of obtaining an assortment in which:
1) a red candy and a purple candy are found at each end of the tube?
2) a red candy or a purple candy is found at one end of the tube?
3) two candies of the same colour are found at each end of the tube?

Companies spend a lot of money designing and manufacturing new types of packaging for their products. In fact, well-designed packaging can: reduce the quantity of packaging material, reduce storing or shipping costs, decrease the space or weight of the packaged good, make a product more attractive and provide additional protection.

9 How many different ways are there of placing the marbles in each of the following games?

a)

b)

10 The following items are featured on a restaurant menu.

a) How many different meals can be created if only one main dish, one side dish and one type of vegetable are chosen?

b) If one meal is chosen at random, what is the probability that it:

1) consist of a chicken breast with rice and mixed vegetables?

2) consist of ribs with fries in the same meal?

Main dish	Side dish
Chicken breast	Fries
Chicken leg	Boiled potatoes
Ribs	Baked potato
Combo	Rice

Vegetable
Mixed vegetables
Cooked carrots
Green peas

11 A group of 16 children is separated into two soccer teams each with the same number of players.

a) How many different teams can be formed?

b) How many different ways can the shirts, numbered from 1 to 8, be distributed within one team?

12 Within a group of 30 athletes, one athlete is chosen at random to run, another athlete is chosen at random to cycle, another athlete is chosen at random to swim and another athlete is chosen at random to row. How many possible choices are there to complete a quadrathlon?

A quadrathlon is a race in which participants must complete a non-stop series of swimming, rowing, cycling and running. Not only must the participants excel in each sport, but they must also be able to transition quickly from one sport to the next since the time is not stopped once the race starts.

13 An access code is made up of 6 capital letters. A letter can be used more than once:

a) What is the probability of guessing the access code at random on the first try?

b) If a number replaces one of the letters of the access code, does the level of security increase or decrease? Explain your answer.

14 For an electronics project, Karl must place 4 light-emitting diodes next to each other. The diodes are red, green, yellow and blue.

a) In how many different ways can Karl position his diodes?

b) Once his project is completed, the diodes are to light up one after the other in a random way. How many different ways can this arrangement of diodes light up?

A light-emitting diode (LED) is an electronic component capable of emitting light. These components have many applications.

15 A random-sample survey is being conducted at Maryssa's school. Considering that the sample is made of 128 students and the school consists of 1341 students, what is the probability that Maryssa is:

a) the last student chosen to be part of the sample?

b) one of the students chosen to be part of the sample?

c) not chosen to be part of the sample?

16 A CD player randomly selects 10 different songs on a CD that has 16 songs.

a) How many different selections can there be?

b) What is the probability that:

1) song number 5 is chosen first?

2) song number 8 is not chosen at all?

At the end of the 1980s before compact discs were available, music was recorded on vinyl records. Being more fragile and less manageable, vinyl records were quickly replaced by compact discs. In spite of this, some people still prefer the quality of sound provided by vinyl records.

17 **AREA CODES** In Québec, during 2008, there were 5 regional area codes: 819, 418, 450, 514 and 438. The number that follows the area code cannot be a 0 or a 1.

a) How many possible phone numbers were there in Québec in 2008?

b) Three people, whose phone numbers have different area codes, meet. Aside from their area codes, what is the probability that these three people have the same phone number?

18 **REGISTRATION IN QUÉBEC** Maxime wants to register her new car. What is the probability that her license plate will say "MAX," considering that the plate must have three letters followed by three numbers or three numbers followed by three letters; that letters O, I and U are not used; and that the letters and numbers are chosen at random and that repetitions are allowed?

19 Each of the six illustrated coins were chosen successively and randomly in order to form an arrangement. The heads or tails side of the coin was also randomly chosen.

a) If you only take into account the value of the coins, how many possible arrangements are there?

b) If you only take into account the side of the coin that is showing, how many possible arrangements are there?

c) If you take into account both the value of the coins and the side of the coin that is showing, how many possible arrangements are there?

d) What is the probability that:
   1) the toonie is in fifth position?
   2) the coins situated at both ends of the series are showing heads?
   3) the first coin is a nickel showing tails?
   4) the coins are in ascending order of value?
   5) all the coins are showing the same side?

20 **G8** The Group of Eight (G8) is an international forum of discussion and economic partnership formed by eight of the most economically powerful states and countries of the world: Canada, France, Germany, Italy, Japan, Russia, the United Kingdom and the United States. The European Union is also represented within the G8.

a) If the nine representatives are placed next to each other for an official photo:
   1) how many different ways can they be placed?
   2) in how many arrangements is the Prime Minister of Canada next to the Prime Minister of Japan?

b) How many different ways can the nine flags of the G8 countries be placed if the flags are formed in a circle?

21 Eight schools sent one student each to participate in a debate competition. For the first one-on-one round of debate, two opposing students were chosen at random. What is the probability that School **A** came up against School **B** in the first one-on-one round of debate?

22 During a card game, a deck of 52 cards is shuffled and cards are dealt to four players.

a) How many different hands can be created if:

1) each player receives 6 cards?

2) all the cards are dealt?

b) If both jokers are added, how many different hands can be created:

1) if each player receives 6 cards?

2) if all the cards are distributed except 2?

23 The adjacent picture is of a combination lock.

a) How many different combinations of three different numbers are possible of this type of lock?

b) In the mathematical sense, is the word "combination" used correctly? Explain your answer.

24 If 5 boys and 5 girls are randomly placed in a single file, what is the probability that they are placed in the following order: BGBGBGBGBG?

25 **SUDOKU** In its most common form, Sudoku is composed of a grid containing 9 squares by 9 squares, separated in sub-grids of 3 squares by 3 squares. The objective of the game is to fill up the grid using numbers 1 to 9 without entering the same number more than once on a line, in a column or even in a sub-grid.

How many different ways can numbers 1 to 9 be placed:

a) on the same line?

b) in the same column?

c) in the same sub-grid?

In Sudoku, numbers are only used as a formality. Other symbols, such as letters or little images can be used as well. The rules and object of the game stay the same.

PROBLEM Getting out of chores

Five siblings take turns doing household chores by each tossing a coin.

- If all the coins land on the same side, everyone tosses again.

- If only one of the children gets a different result from the rest of the four, he or she gets out of doing the chore.

- If two children get the same result and the three others another result, the two who have a different result get out of doing the chore.

 Is this method of deciding who has to do the chores fair for everyone?

ACTIVITY 1 The Grey Cup

Below are commentaries made by some sports analysts on a game between the Montréal Alouettes and the Saskatchewan Roughriders.

> The Alouettes have a good chance of winning the Cup. I put them as 3 to 2 favourites.

> I think both teams have an equal chance of winning the Cup.

> The probability of the Roughriders winning is $\frac{2}{5}$.

① ② ③ ④ ⑤ ⑥

> The Roughriders have 3 to 1 odds of losing.

> The probability of the Roughriders winning is 40%.

> The Alouettes have a 75% probability of winning.

a. Fill in the following table.

Prediction for the Alouettes vs. Roughriders game

	Analyst ❶	Analyst ❷	Analyst ❸	Analyst ❹	Analyst ❺	Analyst ❻
Probability of an Alouettes victory						
Probability of an Alouettes defeat						
Odds for an Alouettes victory	3:2					
Odds against an Alouettes victory	2:3					

b. Of the characters depicted above, which ones are saying the same thing?

c. What is the difference between odds and probability?

The Canadian Football League (CFL), officially founded in 1958, consists of 8 teams separated into 2 divisions (East and West) with 4 teams each. Each year, the Grey Cup is awarded over to the champion team of the CFL.

In many North American cities, February 2 is Groundhog Day. Celebrations are based on the following tradition: if a groundhod comes out of its hole and doesn't see its shadow because it's cloudy, then winter is almost at an end. However if the groundhog sees its shadow because it's a sunny day, then the groundhog goes back into hibernation and there will be six more weeks of winter

Based on the *Canadian Encyclopedia*, an analysis of meteorological data of the past 30 to 40 years for 13 Canadian cities shows a 50% probability that February 2 be a sunny day. The same observations have been used to calculate the number of times the groundhog was correct in its prediction.

a. What is the mean percentage of years that the groundhog was right?

b. Complete the probability tree diagram showing that the predictions are accurate by chance.

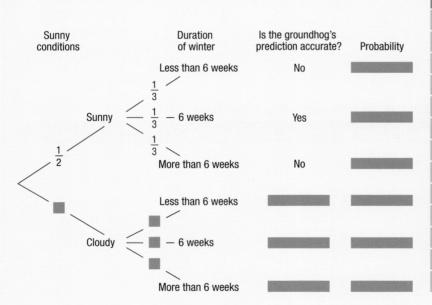

City	Percentage of the years that the groundhog was correct
St. John's	41
Charlottetown	41
Halifax	42
Fredericton	34
Montréal	36
Toronto	29
Ottawa	42
Winnipeg	30
Regina	38
Edmonton	26
Vancouver	35
Whitehorse	42
Yellowknife	50

c. The organizers of Groundhog Day assess that the accuracy level of predictions made by these rodents is 75% to 90%. Is the accuracy of these predictions:

1) calculated by mathematical reasoning?

2) based on a judgment or perception?

d. According to you, are the groundhog's predictions based on chance or are they based on something else? Explain your answer.

Techno math

Graphing calculators allow you to simulate random experiments.

Screen 1

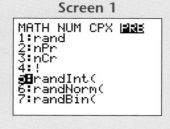

The fifth choice on this screen allows you to randomly generate a series of whole numbers.

This screen simulates tossing a coin ten times: "0" represents tails and "1" represents heads. It is then possible to organize the outcomes in a list.

Screen 2

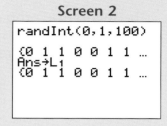

It is possible to sort the outcomes of a random experiment in ascending or descending order.

Screen 3

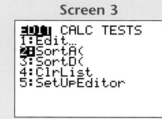

Screen 4

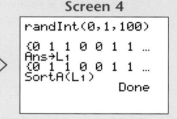

This screen shows the reorganized list in ascending order from which you can observe the frequency of each outcome. You may, among other things, use the frequency to calculate the experimental probability.

Screen 5

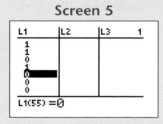

a. Referring to Screen **2**, determine the three numbers that you would have to use with the function `randInt` to simulate the outcomes for:

   1) 25 tosses of a coin

   2) 35 tosses of a 6-sided die, numbered from 1 to 6

   3) 200 tosses of a 6-sided die, numbered from 1 to 12

b. Referring to Screen **5**, what is the experimental probability of obtaining:

   1) tails?   2) heads?

c. 1) After having simulated the results of 50 draws with replacement from a jar containing 8 balls numbered from 1 to 8 on your calculator, determine the experimental probability of drawing ball number 5.

   2) After having simulated tossing a 6-sided die many times with your calculator, verify the following conjecture: The greater the number of experiments performed, the closer the experimental probability is to the theoretical probability.

knowledge 6.2

THEORETICAL PROBABILITY

The **theoretical probability** of an event is a number that quantifies the possibility that this event will occur. This number is determined by using **mathematical reasoning**. In the case when all outcomes are equally probable, the theoretical probability is calculated as follows.

$$\text{Theoretical probability} = \frac{\text{number of favourable outcomes}}{\text{number of possible outcomes}}$$

E.g. When you draw a marble from a bag containing 3 yellow marbles, 2 blue marbles and 8 red marbles, the theoretical probability of the event "drawing a yellow marble" is calculated as follows:

$$P(\text{drawing a yellow marble}) = \frac{\text{number of yellow marbles}}{\text{total number of marbles}} = \frac{3}{13}$$

EXPERIMENTAL PROBABILITY

The **experimental probability** of an event is obtained following **experimentation**. It is often used when the theoretical probability is impossible to calculate.

$$\text{Experimental probability} = \frac{\text{number of times the expected outcome has occurred}}{\text{number of times the experiment was repeated}}$$

E.g. The experimental probability of tennis players successfully completing on their first serve is established based on their performance from previous matches.

The greater the number of repetitions of a random experiment, the more the experimental probability approaches the theoretical probability.

SUBJECTIVE PROBABILITY

The **subjective probability** of an event is based on the **judgment** or **perception** held by a person who has a certain set of information about the situation or the random experiment.

E.g. 1) There is a 60% probability of showers tomorrow.

2) There is a $\frac{1}{3030}$ to $\frac{1}{2700}$ probability that Asteroid 2007 VK187 will collide with the Earth between 2048 and 2057.

152   VISION 6

ODDS FOR AND ODDS AGAINST

Odds for	Odds against
The **odds for** the results of an expected outcome are shown by the relation of the number of possible desirable outcomes to the number of possible undesirable outcomes. $$\text{Odds for} = \frac{\text{number of possible desirable outcomes}}{\text{number of possible undesirable outcomes}}$$	The **odds against** the result of an expected outcome are shown by the relation of the number of possible undesirable outcomes to the number of possible desirable outcomes. $$\text{Odds against} = \frac{\text{number of possible undesirable outcomes}}{\text{number of possible desirable outcomes}}$$
E.g. The odds for the local team winning the tournament are 3:2. This means the local team has 3 chances to win and 2 chances to lose.	E.g. The odds against a player who bets on "0" in roulette are 36:1. This means that the player has 36 chances to lose and 1 chance to win.

It should be noted that the odds for the results of an expected outcome refer to the odds for the event occurring.

DISTINCTION BETWEEN ODDS AND PROBABILITY

$$\text{Probability} = \frac{\text{number of possible favourable outcomes}}{\text{number of possible favourable outcomes} + \text{number of possible unfavourable outcomes}}$$

E.g. The odds that Mikka wins the golf tournament are 4 to 3; this means the same thing as the following statements:

- The odds for Mikka winning the golf tournament are 4:3.

- The odds against Mikka winning the golf tournament are 3:4.

- The probability that Mikka wins the golf tournament is $\frac{4}{4+3} = \frac{4}{7}$.

- The probability that Mikka does not win the golf tournament is $\frac{3}{4+3} = \frac{3}{7}$.

The odds for or the odds against are used to determine potential gain in various situations.

E.g. The odds that the local soccer team win the next game are 5:4.

Amount that a person can win if they bet $10 that the local team will win: $$\frac{5}{4} = \frac{\$10}{\text{Amount to win}}$$ The amount to win is $8. Total remittance of funds: $8 + initial bet of $10 = $18.	Amount that a person can win if they bet $10 that the local team will lose: $$\frac{4}{5} = \frac{\text{Amount to win}}{\$10}$$ The amount to win is $12.50. Total remittance of funds: $12.50 + initial bet of $10 = $22.50.

practice 6.2

There are two types of twins: fraternal twins and identical twins. Based on research, the odds of having twins are the same for all women. However, if there are already fraternal twins on the mother's side of the family, the chances are greater that she will have twins as well.

1 Indicate whether the probability is theoretical, experimental or subjective for each of the following cases.

  a) According to Christine, there is a 25% probability of catching a fish in less than one hour in Lake Heron.

  b) There is a $\frac{1}{13}$ probability of picking an ace in a deck of cards.

  c) The odds that a couple has twins is determined by using demographic data.

  d) The probability of meeting a Hollywood star at the grocery store is relatively low.

  e) There is a 65% probability that there will be good weather next Monday.

  f) The probability that you roll two dice and obtain a sum of 7 is $\frac{1}{6}$.

2 The probability that a local basketball team will win the next game is estimated at $\frac{2}{7}$.

  a) What are the odds for the local basketball team winning the next game?

  b) What are the odds against the local basketball team winning the next game?

  c) What is the possible net gain for someone who bets $5 that the team will:

    1) win?   2) lose?

3 Just before a handball game between the Pythons and the Tornadoes, experts estimated a $\frac{5}{8}$ probability that the Tornadoes win.

The odds for a Pythons' victory are 5:3.

No, the odds against a Tornadoes' victory are 3:5.

Which of these two individuals is right? Explain your answer.

4 A women's college hockey team is favoured by 3 to 2 odds to win. What are the potential gains for someone betting $12:

  a) on the favoured team's victory if they were to win?

  b) on the favoured team's defeat if they were to lose?

5 According to this week's weather forecast, there is a 60% probability of rain for Saturday and a 70% probability of rain for Sunday. What is the probability that it:

a) rains only Saturday?

b) rains Saturday or Sunday?

c) rains Saturday and Sunday?

d) doesn't rain Saturday or Sunday?

Environment Canada defines the probability of showers as the probability of a measurable quantity (0.2 mm or more) of rain in any location in the forecast region during the forecast period.

6 Hanna states that if the probability of showers for a given day is 100%, meaning it will rain all day. Do you agree with her statement? Explain why.

7 If there is a 78% probability of an athlete winning a ping-pong tournament, what are the odds for this athlete's victory?

8 The odds for an event occurring are 1:1.

a) What does this mean?

b) Give an example of a situation where the odds for an event occurring are 1:1.

9 Two dice numbered from 1 to 6 are rolled. What are:

a) the odds for obtaining a sum of 7?

b) the odds against obtaining a sum of 12?

c) the odds for obtaining a sum of 2?

d) the odds for obtaining a sum of 1?

10 What is the probability:

a) of getting a 4 upon rolling a die?

b) that you speak to your best friend today?

c) that you take the bus tomorrow?

The most common dice are cubed, but others are shaped as polyhedrons. A die's surface can consist of numbers, pictures, colours or symbols, depending on its use.

11 Nathan is looking to invest an amount of money for a period of 10 years. The following is a list of suggested investment choices:

Investment A	Investment B	Investment C	Investment D
The odds are 3 to 2 that after 10 years, the value of the investment be at least 1.5 times the initial value.	The probability that after 10 years, the value of the investment be at least 1.5 times the initial value is 55%.	The odds against an increase of at least 50% of the initial value of the investment after 10 years are 1:4.	There are as many odds for as there are odds against that the value of the investment be at least $\frac{3}{2}$ of the initial value after 10 years.

Which one of the four investments should Nathan choose? Explain why.

12 A jar contains 4 blue marbles and 3 white marbles. One marble is drawn from the jar at random and its colour is noted. Determine:

a) the probability of drawing a blue marble

b) the odds for that the outcome is blue

c) the probability of drawing a white marble

d) the odds against that the outcome is blue

13 To determine the following probabilities, specify which approach seems to be the most appropriate (theoretical, experimental or subjective). Explain your answer.

a) the probability of winning a lottery prize

b) the probability of being a victim of a criminal act

c) the probability of a thumbtack falling on a person's head

d) the probability of obtaining tails after tossing a coin

e) the probability of finding a four-leaf clover

What is known as a four-leaf clover is really a clover leaf composed of four leaflets instead of three. Some people think that these rare four-leaf clovers bring luck. Those who search and find four-leaf clovers are called *quatrefoilists* and those who collect them are called *ultraquatrefoilists*.

14 During a handball tournament, a men's team is favoured 5 to 3 to win the tournament:

a) 1) What is the probability that the favoured team wins the tournament?

2) Is this probability theoretical, experimental or subjective?

b) A person bets $15 on the favoured team. What will her net gain be if:

1) she bets that the favoured team is going to win the tournament and she wins?

2) she bets that the favoured team is going to lose and she loses?

15 Match each statement in the left column to a probability in the right column.

Statement	Probability
A The Sphinx has 1 to 2 odds of beating the opposing team.	**1** The probability of a Sphinx victory is 100%.
B The two teams are battling at equal odds.	**2** The probability of a Sphinx defeat is 40%.
C The Sphinx is favoured 3 to 2.	**3** The probability of a Sphinx defeat is 0.5.
D The opposing team has no chance of beating the Sphinx.	**4** The probability of a Sphinx victory is $\frac{1}{3}$.

16 A bag contains 5 green marbles and 3 blue marbles. Three marbles are drawn successively without replacement.

a) What is the probability:

　1) that the first marble drawn is blue?

　2) that the second marble drawn is green, considering that the first marble was blue?

b) What are the odds:

　1) that the first marble drawn is blue?

　2) that the second marble drawn is green, considering that the first marble was blue?

17 When a die is loaded, the probability of obtaining an outcome is no longer the same for all the sides of the die. The following outcomes are bases on 1000 throws of a loaded die.

a) 1) For each side, calculate the probability of obtaining that outcome.

　2) Is this a theoretical, experimental or subjective probability?

b) Suggest a way of detecting whether a die is loaded or not.

Side	Number of times this side was obtained
1	130
2	132
3	128
4	135
5	345
6	130

18 Martine's soccer team is participating in a regional tournament. Since there are 12 teams participating in the same tournament, Martine states that the odds for her team winning is 1 to 11. Explain why Martine could be wrong.

19 **GALTON BOARD** A Galton board is a device consisting of an inclined board and staggered rows of nails. Balls are dropped at the top of the board and pass either to the right or the left of whatever nail it touches. A series of bins are laid out at the bottom of the board in order to collect the balls. Suppose 1000 balls are dropped onto the Galton board, as shown in the illustration below.

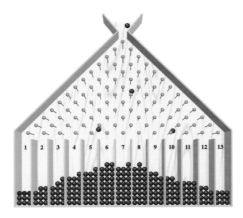

a) Before starting the experiment, it is concluded that the probability of a ball hitting a nail and falling to the right side is $\frac{1}{2}$. Would this probability be theoretical, experimental or subjective? Explain your answer.

b) Is it possible to calculate the exact number of balls that will be collected in Bin **4** before the experiment? Explain your answer.

c) Is it possible to calculate the probability that a ball will land in Bin **4** before the experiment? Explain your answer.

d) After the balls drop, 30 balls are found in Bin **4**. This data is then used to calculate the probability of a ball collect in Bin **4**. Is this probability theoretical, experimental or subjective? Why?

20 **LIGHTNING** In Canada, lightning strikes approximately 2.7 million times a year.

a) Considering that the Canadian population is 33 140 000, can the probability of getting struck by lightning be calculated? Explain your answer.

b) Each year in Canada, lightning kills approximately 6 people and seriously injures approximately 70 people. Can the probability of getting struck by lightning be calculated with this information? Explain your answer.

Lightning is a natural phenomenon consisting of an electrical discharge caused by an accumulation of electrical charges in the atmosphere followed. This is by electrostatic reequilibration which causes a bright light (lightning) and a strong explosion (thunder).

21 The Amiral volleyball team is favoured 3 to 2 against its rival Les Riverains. In the playoffs, the two teams play up to 7 games against each other. The team that wins 4 of these games goes on to the next round while the other team gets eliminated. What is the probability that Les Riverains get eliminated by Amiral in 5 games?

Volleyball was invented in the United States at the end of the 19th century by a physical education teacher. In those days, the number of players was not defined, and the rules were very different from the ones that exist today. Nowadays, it is one of the most popular sports in the world: approximately 260 million people play regularly.

22 It is estimated that there is a 50% probability that a baby being born is male.

a) Is this probability theoretical, experimental or subjective? Explain your answer.

b) Complete the following table based on the composition of a family of 4 children.

Composition of the family	Subjective probability	Theoretical probability
B, G, B, G in this order		
2 boys and 2 girls in any order		
4 boys		

23 GEOCACHING Geocaching is an outdoor activity similar to a treasure hunt in which participants use a GPS. The following is a recent example of a geocaching game: "The coordinates of the cache are 45°24 N.XXX 73°29 W.XXX where the Xs are replaced by numbers 0, 1, 3, 5, 6 and 7." What are the odds for finding these coordinates on the first try by randomly replacing the Xs by the given numbers?

24 A bag contains red, yellow, blue, black and white marbles.

- The odds of drawing a white marble are 1 to 14.
- The odds of not drawing a red marble are 13 to 2.
- The odds of drawing a yellow marble are 1 to 4.
- The odds of not drawing a blue marble are 11 to 4.

Considering that the bag contains 60 marbles, how many marbles of each colour are there?

Geocaching consists of hiding sealed containers called *geocaches*. These containers hold small objects. Whoever leaves the geocache reveals its geographic coordinates so that the other players can try to find it by using a GPS. Several hundreds of millions of geocaches can be found in over 220 countries worldwide.

This section is related to LES 12 and 13.

PROBLEM   A housing project

A company that specializes in the construction of houses is projecting that it will be constructing many houses in a particular region. A study finds that this company makes, on average, a profit of $15,000 for each house built. The type of soil that the house is built on will affect the profit due to additional costs. The table below summarizes this information.

Results of study

Type of soil	Probability that a house be built on this type of soil	Additional costs ($)
Limestone	20%	5,000
Clay	15%	20,000
Silt	50%	15,000
Sand	15%	8,000

Soil is classified by the particles it is composed of. Clay, for example, is composed of particles whose size is less than 0.002 mm.

Based on the results, should the company undertake this project?

Constructing on clay soil can pose various problems. Clay is, by far, the most frequently used construction material in the world.

ACTIVITY 1 Investment strategies

In the stock market, high-risk investments offer a more significant potential gain. On the other hand, low-risk investments often offer a weaker return. It is for this reason that investment advisors recommend that investors diversify their investments.

The biggest stock market is the New York Stock Exchange often referred to as "Wall Street." This is where the 1929 stock market crash occured. This stock crisis marked the beginning of the Great Depression, a major economic crisis of the 20th century.

Leo wants to invest $2,000. An investment advisor presents the following choices.

Profits generated by an investment of $2,000 after 1 year

Portfolio A		Portfolio B		Portfolio C		Portfolio D	
Profit ($)	Probability (%)	Profit ($)	Probability (%)	Profit ($)	Probability (%)	Profit ($)	Probability (%)
2,000	10	2,000	20	2,000	30	2,000	40
1,500	30	1,500	10	1,500	20	1,500	10
1,000	40	1,000	40	1,000	0	1,000	0
0	15	0	20	0	0	0	10
-1,000	5	-1,000	10	-1,000	50	-1,000	40

a. In Portfolio **A**, do all 5 profits have the same probability of being generated?

b. For each portfolio, calculate the mean annual profit that Leo can expect to earn on a long-term basis.

c. Which portfolio should Leo choose?

d. Complete the following table so that the mean annual profit of Portfolio **E** is $640.

Portfolio E	
Profit ($)	Probability (%)
2,000	20
1,500	20
0	10
-1,000	30

Gabrielle is playing on a seesaw with her father. Because her father is a lot heavier than she, the seesaw leans toward her father's side.

If two of Gabrielle's friends come join her or if the pivot point of the seesaw moves, the seesaw leans toward's Gabrielle's side.

If only one friend joins Gabrielle or if the pivot point is moved again, the seesaw is balanced.

The illustrations below represent a situation where potential gains and losses are being weighed against the probability of a gain or a loss occurring.

P_1: Probability of winning

P_2: Probability of losing

Situation A

$P_1 = 25\%$ $P_2 = 75\%$ $40 $10

Situation B

$P_1 = 25\%$ $P_2 = 75\%$ $30 $10

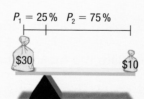

Situation C

$P_1 = 25\%$ $P_2 = 75\%$ $25 $10

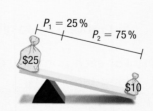

a. Which one of the following situations can be considered:

1) advantageous?   2) disadvantageous?   3) fair?

b. Which modifications can be made to a disadvantageous situation to make it fair if the modification would only be made to:

1) the potential loss?

2) the potential gain?

3) the probability that a gain occurs?

4) the probability that a loss occurs?

Techno math

A spreadsheet allows you to calculate the mathematical expectation of data entered into cells. For example, it is possible to determine expected gain for a random experiment.

The gain and the probability of realizing this gain are entered into cells.

Formulas allow you to automatically generate calculations on the data of the experiment.

Screen 1

	A	B	C	
1	Gain ($)	Probability P	Gain × P ($)	
2	20	0.1	2.00	=A2•B2
3	50	0.05	2.50	=A3•B3
4	-15	0.75	-11.25	=A4•B4
5	100	0.01	1.00	=A5•B5
6	65	0.02	1.30	=A6•B6
7	25	0.03	0.75	=A7•B7
8	35	0.04	1.40	=A8•B8
9				
10	Total:	1		=sum(B2:B8)
11	Expected gain ($):		-2.30	=sum(C2:C8)

By changing one or more data, you can observe that various results are automatically modified.

Screen 2

	A	B	C
1	Gain ($)	Probability P	Gain × P ($)
2	20	0.1	2.00
3	50	0.05	2.50
4	-20	0.75	-15.00
5	100	0.01	1.00
6	65	0.02	1.30
7	25	0.03	0.75
8	35	0.04	1.40
9			
10	Total:	1	
11	Expected gain ($):		-6.05

Screen 3

	A	B	C
1	Gain ($)	Probability P	Gain × P ($)
2	20	0.01	0.20
3	50	0.05	2.50
4	-15	0.75	-11.25
5	100	0.1	10.00
6	65	0.02	1.30
7	25	0.03	0.75
8	35	0.04	1.40
9			
10	Total:	1	
11	Expected gain ($):		4.90

a. Referring to Screen **1**, what changes have been made to:

1) Screen **2**?   2) Screen **3**?

b. Explain what the products shown in Column **C** represent.

c. Think of and explain an experiment or a game that could be represented by the data shown in Screen **1**.

d. Cell B10 allows you to verify an important property when referring to expected gain. What is it?

e. Reproduce Screen **1** using a spreadsheet. Modify the data to create a random experiment that is the most fair by modifying only the data in:

1) Column **A**   2) Column **B**

knowledge 6.3

MATHEMATICAL EXPECTATION

During a random experiment, the **mathematical expectation** corresponds to the sum of the products of the values of a random variable and its probability. Thus, the mathematical expectation corresponds, in some way, to the mean of the values of the random variables weighed by the probability associated to each of these values.

> Mathematical expectation $= p_1 \times r_1 + p_2 \times r_2 + p_3 \times r_3 + \ldots + p_n \times r_n$
>
> where $p_1, p_2, p_3, \ldots, p_n$ correspond to the probabilities of obtaining outcomes 1, 2, 3, ..., n, while $r_1, r_2, r_3, \ldots, r_n$ correspond to the numerical value of the outcomes 1, 2, 3, ..., n.

In the case of a game that has a probable gain and a probable loss, the mathematical expectation is more often referred to as *expected gain* and is represented as:

> Expected gain = (probability of winning) × (net gain) + (probability of losing) × (loss)
>
> (gain) – (initial bet)　　　　　corresponds to the initial bet

E.g.　1)　In a game of roulette, a person betting on a winning number receives 35 times the amount of the bet plus the return of the initial bet. Since the roulette wheel has slots numbered from 0 to 36, the probability of the ball falling on a given number is $\frac{1}{37}$.

The expected gain for a \$10 bet in a game of roulette is therefore:

$$\text{Expected gain} = \frac{1}{37} \times 350 + \frac{36}{37} \times \text{-}10 = \text{-}\frac{10}{37} \text{ or } \sim \text{-} \$0.27.$$

probability of winning　　probability of losing

2)　Let $\Omega = \{1, 2, 3, \text{-}4\}$ and $P(1) = 0.2$, $P(2) = 0.15$, $P(3) = 0.4$ and $P(\text{-}4) = 0.25$:

Mathematical expectation $= 0.2 \times 1 + 0.15 \times 2 + 0.4 \times 3 + 0.25 \times \text{-}4 = 0.7$

FAIRNESS

A game in which the mathematical expectation is greater than zero puts the participants at an advantage. A game in which the mathematical expectation is less than zero puts the participants at a disadvantage. When the mathematical expectation of a game is equal to zero, the game is considered fair.

E.g.　1)　A game consists of randomly drawing a marble from a set composed of 5 red marbles and 4 blue marbles. If a blue marble is drawn, the participant wins \$12.50 in addition to the initial bet. If a red marble is drawn, the initial \$10 bet is lost. This game is fair because:

$$\text{Mathematical expectation} = \frac{4}{9} \times 12.50 + \frac{5}{9} \times \text{-}10 = \$0.$$

2)　For \$2, a person can participate in a draw for a coupon valued at \$50. The probability of winning this draw is $\frac{1}{100}$. This draw is not fair because:

$$\text{Mathematical expectation} = \frac{1}{100} \times 48 + \frac{99}{100} \times \text{-}2 = \text{-}\$1.50.$$

practice 6.3

1 Calculate the mathematical expectation for each of the following situations:

a) $\Omega = \{1, 2, 7, 8\}$;
$P(1) = 15\%$, $P(2) = 20\%$, $P(7) = 25\%$, $P(8) = 40\%$

b) $\Omega = \{-10, -7, 2, 8\}$;
$P(-10) = \frac{1}{10}$, $P(-7) = \frac{3}{20}$, $P(2) = \frac{1}{2}$, $P(8) = \frac{1}{4}$

c) $\Omega = \{-5, 10, 100, 1\,000\,000\}$;
$P(-5) = 99.0991\%$, $P(10) = 0.9\%$, $P(100) = 0.0009\%$,
$P(1\,000\,000) = 0.000\,000\,009\%$

d) $\Omega = \{-5, 10, 15, 20\}$;
$P(-5) = 0.75$, $P(10) = 0.1$, $P(15) = 0.1$, $P(20) = 0.05$

e) $\Omega = \{2.5, 4, 6\}$;
$P(2.5) = \frac{4}{5}$, $P(4) = \frac{1}{10}$, $P(6) = \frac{1}{10}$

2 For each case, determine the value of x for which the mathematical expectation is zero.

a) $\Omega = \{-4, -2, 0, 2, x\}$;
$P(-4) = 20\%$, $P(-2) = 10\%$, $P(0) = 45\%$, $P(2) = 10\%$, $P(x) = 15\%$

b) $\Omega = \{-1000, -800, x, 2000, 40\,000\}$;
$P(-1000) = 0.10$, $P(-800) = 0.15$, $P(x) = 0.17$, $P(2000) = 0.5$, $P(40\,000) = 0.08$

c) $\Omega = \{-3, 22\}$;
$P(-3) = x$, $P(22) = 12\%$

d) $\Omega = \{-6, 18\}$;
$P(-6) = \frac{2}{3}$, $P(18) = x$

3 At a fair, there is a game in which a person must draw a card from a deck of 54 cards. If one of the four aces is drawn, the prize is a box of candy valued at $10. If one of the two jokers is drawn, the prize is a teddy bear valued at $20. If any other card is drawn, there is no prize. If it costs $2 to participate, is this game fair? If not, how much should participation cost so that the game is fair?

The origins of the teddy bear are relatively recent. In Germany, around 1880, Margaret Steiff made teddy bears with leftover fabric and stuffed them with chipped wood or sawdust. At the start of the 20th century, the popularity of these teddy bears spread to the rest of Europe and to North America. This popularity continues to grow, and teddy bears have become the object of affection for children and a passion for collectors.

4 A new lottery game appears on the market. Participation consists of choosing 6 numbers out of 49. Each week, 6 numbers are chosen at random and winnings are given to the participants who chose 2, 3, 4, 5 or 6 of the winning numbers. The prizes are awarded based on the adjacent table.

Number of winning numbers in the selection	Winnings ($)	Probability
6	8,000,000	$\frac{1}{13\ 983\ 816}$
5	1,850	$\frac{1}{55\ 492}$
4	65	$\frac{1}{1033}$
3	10	$\frac{10}{567}$
2	2	$\frac{5}{406}$

a) What is the expected gain of this lottery if the bet is $1?

b) What is the expected gain of this lottery if the bet is $2?

c) What should the initial bet be so that this lottery is fair?

d) Explain why the expected gain from public lotteries is generally less than zero.

5 The owner of a newspaper stand buys a current affairs magazine for $2.35 a copy and resells it for $5.25 a copy. Based on his sales for the last six months, he estimates the probability of selling a certain number of copies of this magazine as follows:

Number of magazines sold	0	1	2	3	4	5	6	7	8	9	10	11	12
Probability	$\frac{1}{4096}$	$\frac{3}{1024}$	$\frac{33}{2048}$	$\frac{55}{1024}$	$\frac{495}{4096}$	$\frac{99}{512}$	$\frac{231}{1024}$	$\frac{99}{512}$	$\frac{495}{4096}$	$\frac{55}{1024}$	$\frac{33}{2048}$	$\frac{3}{1024}$	$\frac{1}{4096}$

a) What is his expected gain if he orders:
  1) 12 copies?
  2) 10 copies?
  3) 6 copies?

b) At what minimum price should he sell his magazines so that his expected gain is positive, if he orders:
  1) 12 copies?
  2) 10 copies?
  3) 6 copies?

According to *l'Institut de la statistique du Québec*, in 2005, approximately 7% of Quebeckers' leisure expenses were spent on newspapers, books and magazines. The mean total for the purchase of electronics, such as audio and television equipment, CDs and DVDs represented approximately 26.4% of leisure expenses.

6 Maude and Jonathan are playing a game of heads or tails. They take turns tossing a loonie. If it lands on heads, Jonathan gives Maude $1. If the coin lands on tails, Maude gives Jonathan $1. The game ends once one of the players has no coins left.

a) For each toss, what is the probability that:

1) Maude wins $1?

2) Maude loses $1?

b) What is the mathematical expectation of this game?

c) Is this game fair?

d) Considering that Maude starts with $1, complete the following table.

Number of tosses	1	2	3	4	5	6	7	8	9	10
Maude's holdings ($)	Probabilty									
11	–	–	–	–	–	–	–	–	–	≈ 0.0010
10	–	–	–	–	–	–	–	–	≈ 0.0020	–
9	–	–	–	–	–	–	–	≈ 0.0039	–	
8	–	–	–	–	–	–	≈ 0.0078	–		–
7	–	–	–	–	–	≈ 0.0156	–		–	
6	–	–	–	–	≈ 0.0313	–		–		–
5	–	–	–	0.0625	–		–		–	
4	–	–	0.125	–		–		–		–
3	–	0.25	–	0.1875	–		–		–	
2	0.5	–	0.25	–		–		–		–
1	–	0.25	–	0.125	–		–		–	
0	0.5	–	0.125	–		–		–		–

e) What is the probability that Maude has no money left after:

1) 1 toss?

2) 2 tosses?

3) 3 tosses?

4) less than 5 tosses?

5) less than 10 tosses?

f) Complete the following table considering that Jonathan starts with $5.

Number of tosses	1	2	3	4	5	6	7	8	9	10
Jonathan's holdings ($)	Probability									
11	–	–	–	–	–		–		–	
10	–	–	–	–		–		–		–
9	–	–	–		–		–		–	
8	–	–		–		–		–	–	–
7	–		–	–	–		–	–	–	–
6		–		–		–		–		–
5	–		–	–	–	–	–	–	–	–
4		–	–	–		–		–		–
3	–		–	–	–	–	–	–	–	
2	–	–		–		–		–		–
1	–	–	–		–	–	–		–	
0	–	–	–	–		–		–		

g) What is the probability that Jonathan has no money left after:
1) 5 tosses?
2) 6 tosses?
3) 7 tosses?
4) less than 5 tosses?
5) less than 10 tosses?

h) Between two players, each holding a different initial amount of money, which one has a greater probability of losing everything? Explain your answer.

i) Between a casino and a person playing at the casino, which one has a greater probability of losing everything? Explain your answer.

Some people develop an addiction to games of chance and money. They develop an uncontrollable urge to bet money: this is known as gambling addiction. In 2002, in Québec, 0.8% of the adult population had a gambling addiction.

7 The probability of the weekly variation of a stock is estimated as follows: 8% probability of a $5 increase, 45% probability of a $4 increase, 20% probability of a $2 increase, 17% probability of a $1 increase and 10% probability of a $2 decrease.

a) Does this stock represent a good investment? Explain your answer.

b) Sebastian wants to buy 180 shares. How much should he expect to gain or lose after 2 years?

8 At a fundraiser, there is a half-and-half draw that allows participants to buy tickets for a fixed amount. The winner is chosen by a random draw and wins half of the total of the ticket sales; the fundraiser gets the other half.

a) If *x* represents the number of tickets sold and *y* represents the cost of a ticket, determine the rule that allows the calculation of the expected gain for this type of draw.

b) Is this draw fair? Explain your answer.

Since 1961, at the start of spring each year, volunteers from the Canadian Cancer Society have been selling daffodils to the public as a means of funding activities, services and research. In 2006, in Québec, two million daffodils were sold in four days; this generated a total of $2.3 million. This activity makes the Canadian Cancer Society the largest daffodil buyer in the world.

9 Two people are playing a game of cards in which the objective is to get a card that has a greater value than the opponent's card. Hugo and Eric decide on the following rules. On every turn, they each bet a token and whoever wins the round takes both tokens. If they both have equal cards, they each take back their bets. Hugo and Eric play until one of them has no tokens left. After each turn, all the cards are collected in order to play each round with a complete deck of 52 cards. It is estimated that there is a 6% probability that one round is nul, and the odds that one or the other wins are equal.

a) What is Hugo's expected gain from this game?

b) Which of the two players will win all of the other player's tokens if from the start, the two players:

1) have the same number of tokens?

2) do not have the same number of tokens?

10 The following game is created: a participant chooses a number from 1 to 6 and rolls a die. If the outcome corresponds to the chosen number, the participant wins a sports car valued at $900,000. If not, he or she loses his or her initial bet. The bet for this game is $100,000.

a) Calculate the expected gain for this game.

b) What is the risk incurred by someone participating in this game?

c) Explain why very few people would participate in this game.

The Ferrari 250 GTO was designed in the early 1960s. Less than 40 models of this high-performance classically-designed race car were built. Since the early 1970s, interested collectors have been responsible for price increases so that price tags now exceed $10 million US.

d) Presume that the cost of participating is lowered to $10,000 and the value of the car is lowered to $500,000.

1) What is the expected gain?

2) What is the risk incurred?

e) Walter concludes that a game in which the mathematical expectation is greater than or equal to zero does not present any risk. Explain why Walter's reasoning is false.

Chronicle of the past

Pierre-Simon de Laplace

His life

Born on March 23, 1749, in Beaumont-en-Auge, Pierre-Simon de Laplace started his studies at the Collège de Beaumont. At the age of 16, he entered Université de Caen where he studied integral calculus, cosmology, mathematical astronomy, cause and effect and the theory of games of chance. After his studies, he became a mathematics professor at l'Ecole Militaire in Paris.

Pierre-Simon de Laplace

Thanks to the generosity of Laplace's wealthy neighbours, Laplace, a farm boy, was able to pursue his education.

His work

Laplace's contributed to many scientific areas including astronomy, physics and mathematics, but primarily probability. He was a member of the Weights and Measures Commission and participated in the creation of the metric system. Two of Laplace's most important works are *The System of the World* and *The Analytic Theory of Probability*.

In *The System of the World*, Laplace deals with the movement of celestial bodies and attempts to diminish the part linked to chance in his observation of natural phenomena.

In *The Analytic Theory of Probability*, Laplace borrows the reasoning of logic by induction. He also reveals his rule of succession: If a single event only presents two possible outcomes, be it "success" and "failure," the probability that the outcome of the next event be "success" is calculated according to the following formula:sion:

$$P(\text{success}) = \frac{s + 1}{n + 2}$$ where s is the number of successes previously observed and n is the total number of attempts.

Laplace's rule of succession was the target of many critics because of the absurd example that Laplace used to represent this rule. He calculated the probability that the Sun would rise the next day even though it had risen every day since the beginning of time.

Thus,

P(the sun will rise tomorrow) $= \dfrac{n+1}{n+2}$ where n is the number of days the sun has risen since the beginning of time.

In his example, Laplace estimated the value of n to be 1 826 313 days. He himself acknowledged the absurdity of his example, knowing very well that the probability of the Sun rising the next day was much greater since it is impossible to prevent this event from happening.

Oh Sun, oh great Sun, will I see Thee tomorrow?

Pierre-Simon de Laplace is among the 72 names of scientists inscribed in tribute by Gustav Eiffel on the frieze of the four sides of the Eiffel Tower. Covered with paint in the early 20th century, their names were restored in 1986 and 1987.

1. After 121 coin tosses, there are 63 tails and 58 heads. Using Laplace's rule of succession, calculate the probability of obtaining tails on the 122nd coin toss.

2. According to Laplace's rule of succession, after 38 successive coin tosses, the probability of obtaining tails on the 39th toss is estimated at 47.5%. How many times did it land on heads during the first 38 tosses?

3. The probability obtained with Laplace's rule of succession is always similar to one of the types of probability (theoretical, experimental or subjective). Which one? Give a possible explanation for this similarity.

4. Using Laplace's rule of succession, calculate the probability that the Sun will rise tomorrow, considering that the age of the Earth is estimated to be around 4.5 billion years.

Financial analysts are professionals who study law, economics, marketing or finance. They mostly work for financial institutions but also for real estate, investment and insurance companies.

The role of a financial analyst consists of evaluating risks and the potential gain of some investments in order to allow the companies and individuals to better manage their assets.

A dollar today is worth more than a dollar tomorrow!

One of the financial principles is *time value*, based on the fact that investors prefer to benefit immediately from their investment rather than later. This is why the expected gain, whether it is short or long term, is taken into consideration when examining investment options.

In finance, the formula $PV = \frac{EV}{(1 + r)^n}$ allows one to determine what the present value *(PV)* should be in an investment so that it attains an expected value *(EV)* after n years at a rate r of growth or interest.

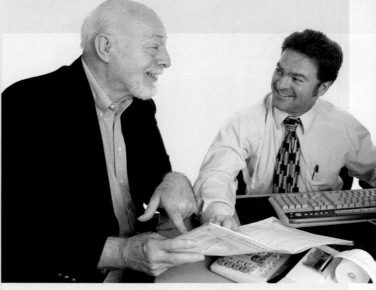

Since the evolution of an investment may vary due to unpredictable factors, a certain amount of uncertainty remains regarding any investment. For this reason, financial analysts resort to probabilistic calculations in their decision-making process.

Risk, speculation, subjective probability and expected gain

Financial analysts must take into account a large number of factors before interpreting market changes. They must then translate their knowledge into language that will allow their clients to understand the information.

> Because of the political situation in the Middle East, there is an 85% probability that the cost of a barrel of oil will increase from $35 to $50 by the end of this year.

> The discovery of new oil fields will allow for the increase of oil reserves and the stabilization of the price of oil. I estimate a probability of 50% that the price of oil will not change in 6 months.

> But that oil requires very costly means of extraction! There is a 25% probability that the oil price will increase by 10% by the end of this year.

> With the upsurge of oil prices and the environmental crisis, other sources of energy will be developed which will cause a decline in the demand for oil. There is a 75% probability that the price of oil will decrease by half in 10 years.

Even though many investments present little or no risk, others are accompanied by the probability that the gains be less than what was expected or even that a part of the amount invested be lost. This is why many financial advisors recommend that investors diversify their investments in order to lower the effects of these risks.

1. By referring to the table "Probability of gain or loss after 1 year," answer the following questions:

a) What is the expected gain of each investment?

b) What is the expected gain of an investment having combined, in equal shares, investments:
   1) **A** and **B** 2) **A** and **C** 3) **B** and **C**

c) Based on the results of **a)** and **b)**, determine which situation would be the most advantageous for an investor.

2. How much should an individual invest in order for the value of his investment to reach $10,000 at the end of 10 years:

a) if he chooses Investment **A**?

b) if he chooses Investment **B**?

c) if he chooses Investment **C**?

d) if he combines in equal shares investments:
   1) **A** and **B**? 2) **A** and **C**? 3) **B** and **C**?

Probability of gain or loss at the end of 1 year

Investment A	Investment B	Investment C
50% probability that the value will decrease by 10%.	85% probability that the value will decrease by 10%.	75% probability that the value will decrease by 20%.
50% probability that the value will increase by 20%.	15% probability that the value will increase by 50%.	25% probability that the value will increase by 50%.

1 Twelve people are lined up side by side for a group photo. How many different ways can these 12 people be arranged?

2 There will be 3 draws from a bag containing 6 red marbles, 8 yellow marbles and 4 green marbles.

a) What is the probability of randomly drawing a red marble, followed by a yellow marble, followed by a green marble:

1) if the marble is put back into the bag each time?

2) if the marble is not put back into the bag each time?

b) What is the probability of randomly drawing a red marble, yellow marble and green marble if none of the marbles are put back into the bag?

c) What are the odds for the outcome "draw at random a marble of each colour":

1) if the marbles are put back into the bag each time and if the order is taken into account?

2) if the marbles are not put back into the bag each time and if the order is not taken into account?

3) if the marbles are not put back into the bag each time and if the order is taken into account?

3 State whether the following probabilities are theoretical, experimental or subjective.

a) There is a 25% probability of a stock market crash in the next 5 years.

b) The probability of winning this lottery is 1 out of 13 000 000.

c) Based on the samples collected from this forest, the probability that a tree will be infected is 0.48.

d) If the trend continues, there is a 99% probability that this athlete will win a gold medal.

e) I have no chance of finding my keys.

f) If I pick a card at random, the probability that it will be an ace is 1 out of 13.

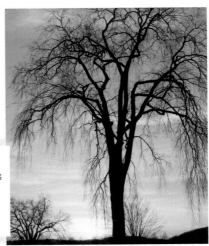

In the early 1930s, a disease called "Dutch Elm Disease" appeared in America and most likely was caused by contaminated wood imported from Europe. In certain zones of Québec, the Elm is almost completely gone. Elsewhere, there are preventative programs in the works to protect the Elm trees that are still intact and to preserve other indigenous species.

4 In a household appliance factory, serial numbers are composed of a series of 2 letters followed by 5 numbers. These serial numbers are engraved on each appliance. All the letters of the alphabet and all numbers from 0 to 9 can be used.

How many different serial numbers can be formed if:

> Serial number: A B 1 2 3 4 5

a) the same letter and number can be used more than once?

b) the same letter can be used more than once but not the same number?

c) the same number can be used more than once but not the same letter?

d) the same number or letter cannot be used more than once?

5 In a fruit juice packaging factory, when a carton is discarded because it was badly packaged, there is a loss of $1.50. A carton of juice can be sold in bulk with a profit of $0.30 or retail with a profit of $0.50. There is a 10% probability that a carton be discarded. Bulk sales are 50% of total sales and retail sales are 40%. What is the mean profit that the company can expect for each carton of juice?

Packaging using cardboard in the shape of a right rectangular prism was commercialized for the first time in 1963. This type of packaging allows better storage of merchandise in a given space than cylinder-shaped packaging such as bottles. This cuts storage and shipment costs. However, this type of packaging is difficult to recycle.

6 The adjacent table provides information on a sample of various bulbs of the same model.

a) Complete the adjacent table.

b) What type of probability is this? Explain your answer.

c) Calculate the mathematical expectation for this situation.

d) Interpret the mathematical expectation in this context.

Durability of bulbs

Number of days after which the bulb stopped working	Number of bulbs	Probability that a bulb will not work during this interval of time
[10, 20[	12	
[20, 30[	50	
[30, 40[	47	
[40, 50[	18	
[50, 60[	76	
[60, 70[	83	
[70, 80[	141	
[80, 90[	168	
[90, 100[	213	
[100, 110[	122	
[110, 120[	74	
[120, 130[	13	
[130, 140[	24	
[140, 150[	9	

7 **BINARY SYSTEM** The binary system is a numeral system whose base is 2. All computer systems use the binary system. The binary system consists only of numbers 0 and 1.

Numbers

Binary system	0000	0001	0010	0011	0100	0101	0110	0111	1000	1001	1010	1011	1100	1101	1110	1111
Decimal system	0	1	2	3	4	5	6	7	8	9	10	11	12	13	14	15

a) If 4 marbles are randomly taken out of a bag containing an equal number of marbles numbered 0 or 1, what is the probability of drawing numbered marbles in the following order 0, 1, 0 and 1?

b) If 1 marble is taken out of a bag containing 16 marbles numbered from 0 to 15, what is the probability of drawing marble number 5?

The first programmable electronic computing device, called the Colossus, was built in England during World War II to decipher encrypted messages in the Lorenz code used between German leaders. Being very large but very quick, the Colossus performed calculations in a few hours that would have normally taken weeks. About ten progressively more efficient versions of the Colossus were built and then destroyed at the end of the war.

8 Ten teams participate in an orientation rally in which everyone leaves from the same point of departure and arrives at the same point. However, each team's route is different and is determined by a draw where the captain must successively pick 12 tokens with the letters that identify the team's intermediate points. The draw is done without replacement and the order in which tokens are chosen determines the route to be followed by the team. What is the probability that:

a) three teams follow the same route?

b) two teams follow the same route?

c) all ten teams follow the same route?

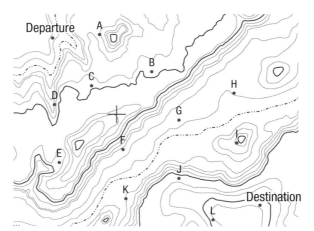

9 The following is the weather forecast for the weekend (*POP* means "probability of precipitation").

a) What is the probability that it rains at some point:
   1) on Saturday?
   2) on Sunday?
   3) during the weekend?

b) What are:
   1) the odds for no rain on Saturday afternoon?
   2) the odds against rain on Sunday evening?

	Saturday	Sunday
Before noon	POP: 20%	POP: 0%
Afternoon	POP: 60%	POP: 40%
Evening	POP: 60%	POP: 10%

After heavy and prolonged rainfall with a risk of flooding, Environment Canada sends out rain warnings. These warnings advise drivers to check road conditions and avoid flooded roads. A rapidly moving water current measuring 15 cm deep can knock a person down, and a rapid current measuring 60 cm deep can wash away a car.

10 Julie's expecting a baby whose sex she does not yet know. There is an estimated 50% probability that the baby will be a boy, 25% that the baby will have blue eyes and 75% that the baby will have brown eyes. Four friends make a bet and contribute equally to buy a box of chocolates valued at $20. The winner of the bet will receive the box of chocolates.

"I bet it's a girl with blue eyes!"

"I'm betting that it's a boy with brown eyes!"

"I'm in! It's a boy with blue eyes!"

"I bet that it's a girl with brown eyes!"

a) For each participant, calculate the mathematical expectation associated with this situation.

b) Explain what is represented by the results obtained.

11 A newly elected board of directors consists of 7 people. Within this board, 1 president, 1 secretary, 1 treasurer, 1 vice-president and 3 administrators must be appointed. Considering that a person cannot occupy more than one position, how many different combinations of boards of directors can be created?

12 In a survey conducted among 100 gardeners:

- 62 people grow cherry tomatoes
- 59 people grow red tomatoes
- 50 people grow yellow tomatoes
- 35 people grow cherry tomatoes and red tomatoes
- 25 people grow red tomatoes and pink tomatoes
- 30 people grow cherry tomatoes and pink tomatoes
- 15 people grow cherry tomatoes, red tomatoes and pink tomatoes
- 5 people don't grow tomatoes

Every year in August, in the city of Buñol, Spain, tens of thousands of people from around the world gather for the *Tomatina*, a famous festival. More than 100 tons of tomatoes are trucked in and are thrown at participants who then use them as ammunition for a two-hour battle.

Out of these people, what is the probability of randomly choosing a person who grows only:

a) red tomatoes?

b) one variety of tomatoes?

c) two varieties of tomatoes?

13 **TAROT** Tarot is a game that uses 78 cards to play a game of the same name. The Tarot cards are distributed in the following way:

- 21 trump cards numbered 1 to 21
- 14 hearts (ace, 2 to 10, jack, knight, queen, king)
- 14 diamonds (ace, 2 to 10, jack, knight, queen, king)
- 14 spades (ace, 2 to 10, jack, knight, queen, king)
- 14 clubs (ace, 2 to 10, jack, knight, queen, king)
- the joker, a card with an image of a mandolin player

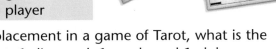

a) If 5 cards are successively picked without replacement in a game of Tarot, what is the probability of obtaining 1 trump card, 1 heart, 1 diamond, 1 spade and 1 club:

1) in this order?

2) in any order?

b) If a 6th card is chosen, what is the probability of getting the joker?

14 Two people meet in the street. What is the probability that these two people:

a) were born on the same day of the week?

b) have the same date of birth except for the year?

c) were born in the same month?

15 There is a draw where participants can win a $200 MP3 player by purchasing one of the 1000 tickets available at $2.

a) What is the expected gain for this draw?

b) To make this draw fair:

1) what should be the number of tickets sold if this were the only value to change?

2) what should be the value of the MP3 player if this were the only value to change?

3) what should be the cost of the ticket if this were the only value to change?

16 PIN TUMBLER LOCK A pin tumbler lock is composed of a cylinder pierced with 5 or 6 holes fitted with pistons and pins numbered 1 to 9. If a lock must have at least 1 pin that is not as long as the others, how many different pin tumbler locks can there be:

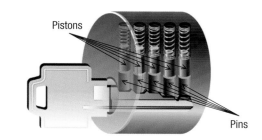

a) if the lock has 5 pins?

b) if the lock has 6 pins?

17 How many different sundaes can be put together if 1 type of topping, 1 type of ice cream and 1 type of decoration are chosen from the following choices?

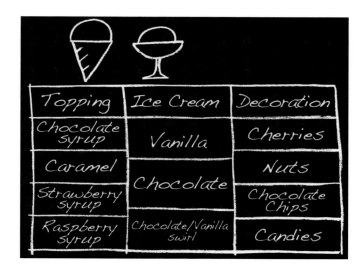

Topping	Ice Cream	Decoration
Chocolate syrup	Vanilla	Cherries
Caramel		Nuts
Strawberry syrup	Chocolate	Chocolate Chips
Raspberry syrup	Chocolate/Vanilla swirl	Candies

18 Four friends are taking the same exam where each friend must choose 5 questions to answer out of a total of 8 questions.

a) How many different combinations of exam questions can be formed?

b) What is the probability that the 4 friends choose the same questions if the questions are chosen at random?

19 During a game show, a contestant is given 3 keys. She must then insert each key in one of the 5 door locks. If she succeeds in putting the right keys in the right locks, the door will open to the next level of the game; if not, she will be eliminated. If the contestant gets 3 tries:

a) what is the probability that she makes it to the next level?

b) are the odds for her going onto the next level greater than the odds against her going onto the next level?

20 THE ST. PETERSBURG PARADOX The St. Petersburg paradox began in the 18th century with Nicolas Bernoulli and was taken over by his cousin, Daniel Bernoulli, a few years later. The St. Petersburg paradox is a game in which a coin is tossed. If the coin lands on heads in the first round, the contestant receives $1, and the game is over. If the coin lands on tails, the coin is tossed a second time. If the coin lands on heads in the second round, the contestant receives $2, and the game is over. If it's tails, the coin is tossed a third time. If the coin lands on heads in the third round, the contestant receives $4, and the game is over. If it's tails, the coin is tossed a fourth time. For each toss, the amount to be won doubles until the coin lands on heads.

a) Complete the following table:

													Total
Number of tosses until heads	1	2	3	4	5	6	7	8	9	10	100	300	–
Winning amount ($)	1	2	4	8	16	32	64	128	256	512			–
Cost of participating ($)	5	5	5	5	5	5	5	5	5	5	5	5	–
Net gain ($)													–
Probability	$\frac{1}{2}$	$\frac{1}{4}$	$\frac{1}{8}$	$\frac{1}{16}$	$\frac{1}{32}$								–
Net gain × probability													

b) Based on the last row of this table, make a conjecture regarding the mathematical expectation of this lottery.

c) Explain how this situation represents a paradox.

21 POETRY Raymond Queneau was a French poet and mathematician. He created a combinatorial work of poetry by writing a 10-page anthology with each anthology separated into 14 horizontal strips. Each strip consists of a different verse. From each line, readers can choose one of the 10 verses to create their own poem.

If two people randomly create their own poem, what is the probability that these two people create the same poem?

bank of problems

22 The owners of a metal-manufacturing company must choose between 3 power presses to use in manufacturing jet-engine parts. The following table shows the performance of these 3 machines.

	Machine 1	Machine 2	Machine 3
Cost of manufacturing the part ($)	150	160	165
Probability that the part is defective (%)	15	10	5

Considering that the manufacturing company sells each part for $300, which machine should the owners purchase?

23 A major department store proposes the following draw: the customer spins the wheel's arrow and receives a 5%, 10%, 20% or 40% discount based on where the arrow stops. To be able to participate in this draw, the customer must make a minimum purchase of $50.

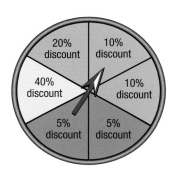

From what initial amount is it advantageous for the customer to add purchases in order to total $50 and benefit from the draw? Explain your answer using probabilistic reasoning.

24 Since there is a 50% probability that a new-born will be a female, Gabrielle claims that a family consisting of 4 children has more chances of being made up of a boy, followed by a girl, followed again by a boy then followed by a girl than simply 4 girls. Explain why this statement is false.

In the United States, the Duggars are a well-known family. Living in a 650-m² house that they built themselves, the Duggars welcomed their 18th child in January 2009. Like all the other children, the first name of the newborn starts with the letter J.

25 To ensure a continuous power supply in the event of a power failure, a hospital is equipped with two generators. The probability of a mechanical failure of both generators is estimated at $\dfrac{1}{200\ 000\ 000}$, and the probability of a mechanical failure of Generator **A** is 2 times greater than a mechanical failure of Generator **B**. What is the probability of a mechanical failure of only one generator?

26 A mining company is evaluating the benefits of an exploration operation. The costs associated with the exploration operations are $3,000,000. If these operations result in finding a gold deposit, the estimated revenue will be $60,250,000. If these operations result in finding a copper deposit, the estimated revenue will be $30,125,000. Geologists estimate a 2% probability of finding a gold deposit and a 4% probability of finding a copper deposit. Determine if the company should undertake this exploration operation.

A gold bar is a small bar composed of pure metal. Its theoretical mass of 1 kg varies between 995 and 1005 g. On the market, each bar is accompanied with a certificate of authenticity. The price of gold, which is listed on the stock exchange, can fluctuate greatly based on the economy. South Africa and Canada mine the most gold.

27 **GENETICS** DNA is comprised of an ordered chain of nitrogen bases whereby the sequence defines a code and allows for the synthesis of protein in cells. Each protein corresponds to a collection of amino acids, and each amino acid is coded by an ordered set of 3 nitrogen bases that repeat. There exist 20 different amino acids and 4 types of nitrogen bases. Using your probabilistic reasoning, explain why amino acids have to be coded by sets of at least 3 nitrogen bases.

The 4 nitrogen bases forming DNA are thymine (T), cystosine (C), guanine (G) and adenine (A). The main function of DNA is to preserve genetic information that defines the development and the functions of living organisms.

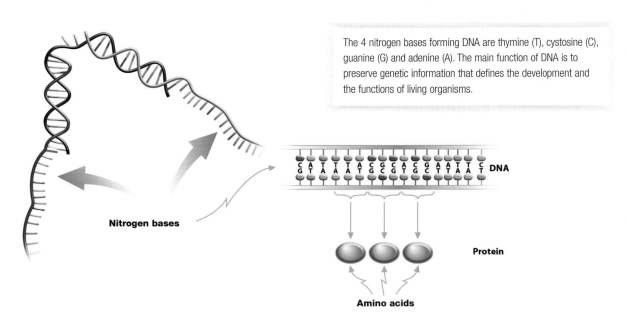

LEARNING AND EVALUATION SITUATIONS

TABLE OF CONTENTS

Personal Finances

In a society where the line between needs and wants is becoming progressively thinner, some people attach a great deal of importance to owning material things and to the consumption of goods and services.

Some people spend more and more and save less and less, thereby worsening their level of debt. Some studies go as far as to indicate that overall, Canadians spend more than they earn.

It is often harder to increase revenue than it is to reduce expenses. Individuals who manage their finances well can easily reduce expenses and balance their personal budget.

This LES is related to sections 4.1 and 4.2.

○ LES 7

Gas savings

A drivers' fuel costs average around $2,000 each year. As the price of gas at the pump keeps rising, many solutions have been suggested to enable drivers to reduce fuel consumption. Not only do these solutions reduce the ecological impact of travel, but they enable drivers to reduce the costs associated with automobile travel.

Proper maintenance and tire pressure, avoidance of sudden manoeuvres and elimination of long idling periods are all ways of reducing fuel consumption. Since over 50% of the energy required to move a vehicle is used to compensate for wind resistance, reducing the vehicle's mean speed also results in fuel savings.

The table below shows the gas consumption for several trips between Montréal and Québec City at a speed exceeding 60 km/h.

Based on the information provided in the adjacent table, you will produce an advertising poster that clearly specifies the financial benefits of reducing mean speed.

Fuel consumption when speed exceeds 60 km/h	
Mean speed exceeding 60 km/h	Increased gas consumption (L)
0	0
15	1.62
21	3.16
27	5.25
35	8.82
43	13.31
52	19.47
56	22.58

This poster has to include the following elements:
- a representation of the situation in the form of a scatter plot
- the curve that best fits this scatter plot
- additional fuel consumption for the Montréal-Québec City trip at speeds of 90 km/h, 100 km/h and 120 km/h
- the fuel savings, calculated in litres and dollars, for the Montréal-Québec City trip when the mean speed is reduced from 120 km/h to 90 km/h

This LES is related to sections 4.1 and 4.3.

O LES 8

C2

Buying on credit

Sometimes an individual needs to make important purchases without having the cash on hand to pay for them right away. When this situation arises, the buyer can use one among several financing options. The table below outlines different financing options.

Standard credit card purchases	Credit card cash advances	Commercial credit card purchasess	Bank-approved line of credit	Lease-purchase
• Annual interest rate: 18%. • Interest rates on the balance are calculated each month. • No interest is calculated for the first 30 days after purchase. • A minimum payment amounting to 3% of the balance or a minimum of $10 (whichever is higher) must be made each month. • The total balance can be paid off at any time without penalty.	• Annual interest rate: 18%. • Interest rates on the balance are calculated each month. • Interest rates are calculated as soon as the cash advance is made. • A minimum payment amounting to 3% of the balance or a minimum of $10 (whichever is higher) must be made each month. • The total balance can be paid off at any time without penalty.	• Annual interest rate: 28%. • Interest rates on the balance are calculated each month. • No interest is calculated for the first 30 days after purchase. • A minimum payment amounting to 3% of the balance or a minimum of $10 (whichever is higher) must be made each month. • The total balance can be paid off at any time without penalty. • The retailer offers a 5% discount on the purchase if the latter is made through the store card.	• Annual interest rate: 6%. • Interest rates on the balance are calculated each month. • Interest rates are calculated as soon as the line of credit is granted. • A minimum payment amounting to 3% of total amount borrowed must be made each month. • The total balance can be paid off at any time without penalty.	• A fixed payment amounting to 6% of the total amount must be made each month. • Merchandise remains the property of the store until the lease is expired. • The lease is for 24 months. • The balance cannot be paid before the lease expires.

For the purchase of a washer-dryer combo whose cost is $2,369.25, including taxes, determine the following for each of the credit options presented above:

• the time required to pay off this purchase if only the minimum is paid each month
• the total amount paid for the washer-dryer combo once the purchase has been paid off

You will then need to choose the best financing option for a person who can afford to pay off this purchase with a maximum payment of $150 each month.

186

This LES is related to sections 4.1 and 4.4.

O LES 9

C1

What plan should I choose?

Many companies offer a variety of telephone services. While the rates for basic service are the same for all subscribers, long-distance plans vary greatly from one company to another.

The following table shows four long-distance plans that may be added to a basic telephone service.

Basic telephone service: $14.95/month
Touch-tone dialling: $2.80/month
9-1-1 service: $0.19/month
Long-distance calls: $0.25/minute, or one of the following plans.

$2.00	$5.00	$12.95	$24.95
Long-distance calls made in Canada and the United States cost $0.10/min.	Long-distance calls made in Canada and the United States cost $0.05/min.	The first 1000 minutes of long-distance calls made in Canada and the United States are free. Each minute in excess of the 1000-min block costs $0.10.	All long-distance calls made in Canada and the United States are free, with no time limit.

The duration of the calls is calculated as follows.

Actual duration (s)	Duration billed (min)
]0, 60]	1
]60, 120]	2
]120, 180]	3
...	...

Which of the plans shown above offers the best value? Justify your answer.

Measurement

Learning context

Be it quantifying the distance between the Sun and the Earth, the height of Mount Everest, the energy expended by an individual during physical activity or the duration of an event, man has always sought to measure his environment. Units of mass, length, time, force or energy enable us to compare magnitudes to one another and to quantify phenomena in our environment.

Measurement is such an important field that there is a science entirely dedicated to it: metrology. Our government has its own department of measurement known as Measurement Canada.

Various quantities are directly measurable; others are not. For instance, the distance from the Earth to the Moon or the diameter of an atom have not been determined with the aid of giant or microscopic tape measures. Thanks to sometimes disarmingly simple mathematical instruments, it has been possible to deduce such measurements indirectly.

The metre, which was officially adopted as the unit of reference for length in 1790, corresponded to the ten millionth part of the length of the Earth's meridian, and was determined through the application of trigonometry.

« What is admirable is not that the star field is so vast, but that man has measured it. »

Anatole France

This LES is related to sections 5.1 and 5.2.

 LES 10

C1

Measuring time

The sundial is one of the first devices invented to measure time. Its construction reflects several mathematical concepts.

As shown in the adjacent illustration, the sundial consists of a style that casts a shadow on a horizontal surface marked with lines indicating the hours of the day from 6 a.m. to 6 p.m. The angle of inclination of the style must equal the latitude of the location where the sundial is used and must point towards true north.

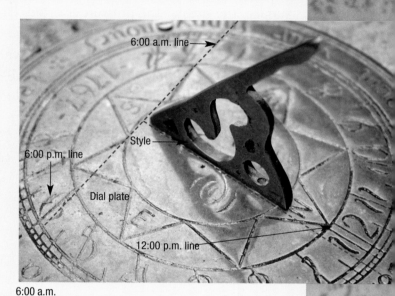

6:00 a.m. line

6:00 p.m. line

Style

Dial plate

12:00 p.m. line

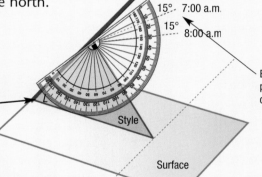

6:00 a.m.
15° 7:00 a.m.
15° 8:00 a.m

The style is shaped like a right triangle whose inclination *L* equals the latitude of the location where the sundial is used.

Style

Surface

Each 15° step on a protractor that is placed perpendicular to the style corresponds to an hour of the day.

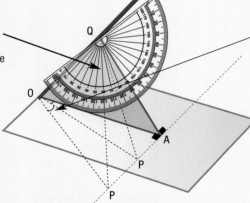

The extension of a line starting from point Q and passing through a given angle determines a point P on the surface of the dial plate.

Q

O

A

P

P

The time line that connects to the base of the style and this point must follow a certain angle.

Determine the angles of each hour line, and create the dial plate of a sundial that could be used in the region where you live.

This LES is related to section 5.3.

LES 11

Measuring pressure

The pressure exerted on a surface by an object is calculated by dividing the mass of the object by the area of the contact surface. For the same mass, the smaller the contact surface, the greater the pressure exerted. On Earth, a 1-kg object exerts a pressure of 9.8 pascals (Pa) on a surface of 1 m².

At a contemporary art show, a sculpture with a mass of 1500 kg is placed on a glass surface that cannot withstand a pressure of more than 16 000 Pa. The show organizers have to choose either one of the two bases shown below.

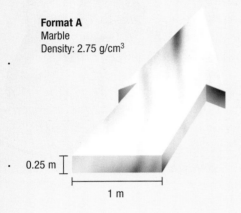

Format A
Marble
Density: 2.75 g/cm³

0.25 m

1 m

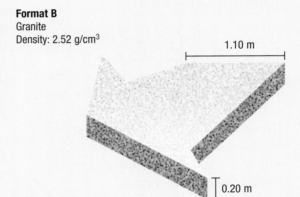

Format B
Granite
Density: 2.52 g/cm³

1.10 m

0.20 m

In order to choose the safest base format, you must do the following:
- Measure the contact surface of the base with the glass surface.
- Calculate the pressure exerted by the sculpture using the formula $P = \frac{9.8\ m}{S}$ where P is the pressure (in Pa), m is the mass (in kg) and S is the contact surface (in m²).

One can easily measure the pressure exerted by a gas or a liquid using a pressure gauge. For large solid objects, one must use an indirect method.

Acquired Immune Deficiency Syndrome

○ Learning context

In January 1983, after a medical follow-up on patients with a deficiency of the immune system, researchers from the Pasteur Institute discovered the human immunodeficiency virus or HIV. This virus attacks the immune system and destroys white blood cells that protect the human body from germs and diseases. An infected human body can no longer defend itself against these foreign bodies and becomes extremely vulnerable to aggressive diseases such as tuberculosis or Kaposi's sarcoma which usually cause death. This is what's called AIDS or acquired immune deficiency syndrome.

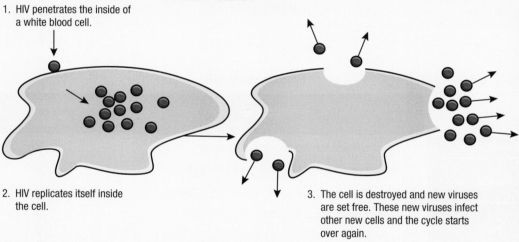

HIV: Chain reaction

1. HIV penetrates the inside of a white blood cell.

2. HIV replicates itself inside the cell.

3. The cell is destroyed and new viruses are set free. These new viruses infect other new cells and the cycle starts over again.

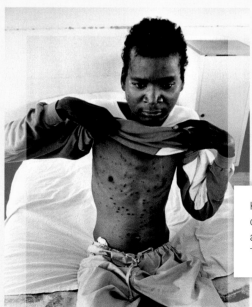

With more than 33 million infected people, AIDS is one of the most serious pandemics known to mankind. This is why the United Nations has made it a priority to fight AIDS. Until now, many treatments have considerably slowed the proliferation of the virus and prevented the disease from evolving to the AIDS stage. Nevertheless, despite all the efforts of the medical profession, a vaccine has not yet been developed.

Kaposi's sarcoma is a disease whereby tumours appear on the skin and certain internal organs such as the lungs, intestines and liver. People who have a weak immune system are much more susceptible to contacting this disease. This is why Kaposi's sarcoma often leads to a diagnosis of AIDS.

This LES is related to sections 6.1, 6.2 and 6.3.

The goal of the treatment is not only to cure people who are already infected with AIDS but to also prevent people who are infected with HIV from developing the disease. Since the introduction of "tritherapy," the death rate linked to AIDS has fallen from 49 000 to 9000 deaths a year in the United States.

LES 12

C1

Polytherapies

In the battle against certain diseases, doctors sometimes resort to polytherapy which consists of treating a disease by combining several medications. Each medication has an effect on a different part of the disease.

Suppose that in order to cure HIV-infected patients, one can choose to combine 1, 2 or 3 types of medication and that these combinations can be made up of medications from the same or different categories. The following is a simulation of the decrease in risk of death related to HIV and the type of treatment used.

Categories and number of medications

Category A	Category B	Category C
Prevents the reproduction of the virus.	Prevents the virus from penetrating the cell of the infected body.	Prevents the new viruses from becoming pathogenic.
5 types of medication	3 types of medication	4 types of medication

Possible treatments

	Distribution	Mean decrease in risk of death (%)
Monotherapy	1 type of medication from one category or another	30
Bitherapy	2 types of medication from the same category	40
	2 types of medication from 2 different categories	55
Tritherapy	3 types of medication from the same category	45
	2 types of medication from one category and 1 type of medication from another	70
	3 types of medication from 3 different categories	82

Situation 1
Each combination of medication has the same chance of being administered.

Situation 2
There is a 75% chance that a tritherapy of medication of the three categories will be administered. The rest of the chances are divided equally among the remaining combinations.

Considering that Situation **2** has priority over Situation **1**, determine how many additional lives can be saved for every 1000 people who would have otherwise died of AIDS.

In the *Millennium Development Goals*, the United
Nations acknowledges the responsibility of the
leaders of the member states towards the world's
most powerless and vulnerable people,
particularly children.

LES 13

C2

Working against HIV

The spread of HIV has devastating effects
on the human condition, particularly in Africa
where in certain countries life expectancy has
dropped significantly.

Leaders from around the world have set certain goals to fight against this major crisis.
The objectives are stated in the United Nations *Millennium Development Goals* and consist
of the following:

- reducing by 25% by the year 2010 the probability that young males and females aged
  15 to 24 years die from AIDS
- reducing by 50% by the year 2010 the probability that children less than 5 years old
  die from AIDS

The following table represents the data linked to mortality in South Africa.

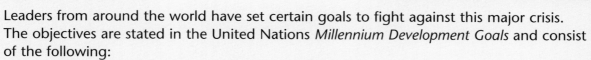

Mortality rate in South Africa

Age	Men		Women	
	Probability of death at this age (%)	Probability that the cause of death be HIV	Number of deaths	Odds for that the cause of death will be HIV
[0, 5[	18.5	$\frac{1}{3}$	45 506	3:5
[5, 15[	1.7	$\frac{2}{17}$	2833	3:13
[15, 25[	6.5	$\frac{1}{19}$	14 826	3:4
[25, 35[	13.6	$\frac{2}{5}$	36 882	7:3
[35, 45[	15.4	$\frac{1}{4}$	28 680	5:4
[45, 55[	12.4	$\frac{1}{4}$	19 946	1:6
[55, 65[	10.4	$\frac{4}{63}$	22 753	1:189
[65, 75[	10.0	$\frac{1}{437}$	27 808	1:5561
[75, 95[	11.5	0	47 765	0
Total	100		246 999	

Statement 1
The objectives of the *Millennium Development Goals* would considerably increase the life
expectancy of a population affected by HIV.

Statement 2
Although the objectives are commendable, more would have to be done so that these interventions
result in a significant impact on the life expectancy of a population affected by HIV.

Using South Africa as a reference, your task is to do the following:
- Calculate the life expectancy for men and women before and after the implementation
  of the objectives of the Millennium.
- Determine which of the above statements are most representative of reality.

REFERENCE
TABLE OF CONTENTS

Graphing calculator

Sample calculations

It is possible to perform scientific calculations and to evaluate both algebraic and logical expressions.

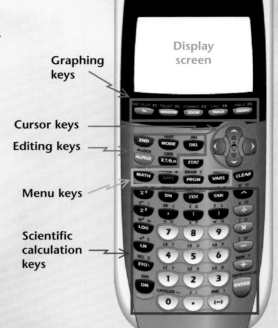

Graphing keys

Display screen

Cursor keys

Editing keys

Menu keys

Scientific calculation keys

Scientific calculations

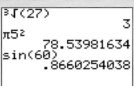

$$\sqrt[3]{(27)}$$
$$3$$
$$\pi 5^2$$
$$78.53981634$$
$$\sin(60)$$
$$.8660254038$$

Logical expressions

$$1/3 = 0.3 \qquad 0$$
$$\sqrt[3]{(216)} = 6 \qquad 1$$
$$6^2 + 7^2 > 8^2 \qquad 1$$

Algebraic expressions

$$5 \rightarrow X \qquad 5$$
$$-2 \rightarrow Y \qquad -2$$
$$5X - 2Y^2 \qquad 17$$

Probability

1. Display the probability menu.

```
MATH NUM CPX PRB
1:rand
2:nPr
3:nCr
4:!
5▮randInt(
6:randNorm(
7:randBin(
```

- Among other things this menu allows the simulation of random experiments. The fifth option generates a series of random whole numbers. Syntax: randInt (minimum value, maximum value, number of repetitions).

2 Display calculations and results.

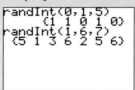

```
randInt(0,1,5)
        {1 1 0 1 0}
randInt(1,6,7)
 {5 1 3 6 2 5 6}
```

- The first example simulates flipping a coin 5 times where 0 represents tails and 1 represents heads. The second example simulates seven rolls of a die with 6 faces.

Display a table of values

1. Define the rules.

```
Plot1 Plot2 Plot3
\Y1▮2^X
\Y2▮0.5X²
\Y3=
\Y4=
\Y5=
\Y6=
\Y7=
```

- This screen allows you to enter and edit the rules for one or more functions where Y is the dependent variable and X is the independent variable.

2. Define the viewing window.

```
TABLE SETUP
 TblStart=0
 ∆Tbl=1
Indpnt: Auto Ask
Depend: Auto Ask
```

- This screen allows you to define the viewing window for a table of values indicating the starting value of X and the step size for the variation of X.

3. Display the table.

X	Y1	Y2
0	1	0
1	2	.5
2	4	2
3	8	4.5
4	16	8
5	32	12.5
6	64	18

X=0

- This screen allows you to display the table of values of the rules defined.

Display a graphical representation

1. Define the rules.

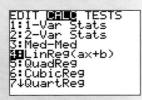

- If desired, the thickness of the curve (normal, thick or dotted) can be adjusted for each rule.

2. Define the viewing window.

- This screen allows you to define the viewing window by limiting the Cartesian plane: Xscl and Yscl correspond to the step value on the respective axes.

3. Display the graph.

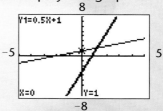

- This screen allows you to display the graphical representation of the rules previously defined. If desired, the cursor can be moved along the curves and the coordinates displayed.

Display a scatter plot and statistical calculations

1. Enter the data.

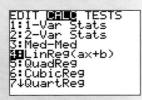

- This screen allows you to enter the data from a distribution. For a two-variable distribution, data entry is done in two columns.

2. Select the mode of representation.

- This screen allows you to choose the type of statistical diagram.

  - ⊡ : scatter plot
  - ⊿ : broken-line graph
  - ▥ : histogram
  - ⊞ : box and whisker plot

3. Display the diagram.

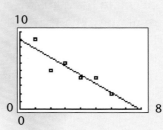

- This screen allows you to display the scatter plot.

4. Perform statistical calculations.

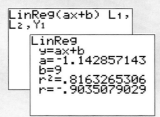

```
EDIT CALC TESTS
1:1-Var Stats
2:2-Var Stats
3:Med-Med
4:LinReg(ax+b)
5:QuadReg
6:CubicReg
7↓QuartReg
```

- This menu allows you to access different statistical calculations, in particular that of the linear regression.

5. Determine the regression and correlation.

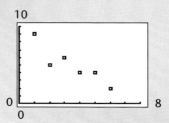

```
LinReg(ax+b) L1,
L2,Y1
LinReg
 y=ax+b
 a=⁻1.142857143
 b=9
 r²=.8163265306
 r=⁻.9035079029
```

- These screens allow you to obtain the equation of the regression line and the value of the correlation coefficient.

6. Display the line.

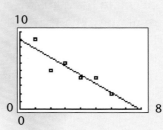

- The regression line can be displayed on the scatter plot.

Spreadsheet

A spreadsheet is software that allows you to perform calculations on numbers entered into cells. It is used mainly to perform calculations on large amounts of data, to construct tables and to draw graphs.

Spreadsheet interface

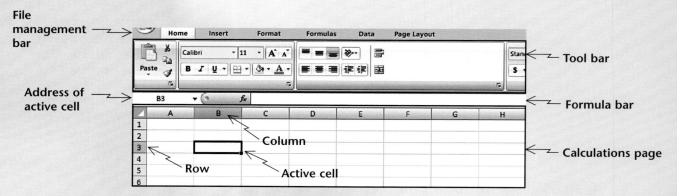

File management bar

Tool bar

Address of active cell

Formula bar

Column

Calculations page

Row

Active cell

What is a cell?

A cell is the intersection of a column and a row. A column is identified by a letter and a row is identified by a number. Thus, the first cell in the upper right hand corner is identified as A1.

Entry of numbers, text and formulas in the cells

You can enter a number, text or a formula in a cell after clicking on it. Formulas allow you to perform calculations on numbers already entered in the cells. To enter a formula in a cell, just select it and begin by entering the "=" symbol.

Example:

Column **A** contains the data to be used in the calculations.

In the spreadsheet, certain functions are predefined to calculate the sum, the minimum, the maximum, the mode, the median, the mean and the mean deviation of a set of data.

	A	B	C	
1	Results			
2	27.4	Number of data	17	=COUNT(A2:A18)
3	30.15			
4	15	Sum	527	=SUM(A2:A18)
5	33.8			
6	12.3	Minimum	12.3	=MIN(A2:A18)
7	52.6			
8	28.75	Maximum	52.6	=MAX(A2:A18)
9	38.25			
10	21.8	Mode	33.8	=MODE(A2:A18)
11	35			
12	29.5	Median	30.15	=MEDIAN(A2:A18)
13	27.55			
14	33.8	Average	31	=AVERAGE(A2:A18)
15	15			
16	33.8	Mean deviation	8.417647059	=MEAN DEVIATION (A2:A18)
17	50			
18	42.3	Standard deviation	11.2543325	=STANDARD DEVIATION (A2:A18)
19				

How to construct a graph
Below is a procedure for drawing a graph using a spreadsheet.

1) Select the range of data.

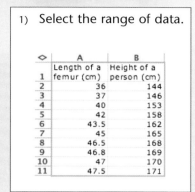

2) Select from the graph assistant.

Cells...
Rows
Columns
Worksheet
Chart...

List...

Page Break
Function...
Name ▶
Comment

Picture ▶
Movie...
Object...
Hyperlink... ⌘K

3) Choose the graph type.

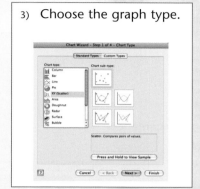

4) Confirm the data for the graph.

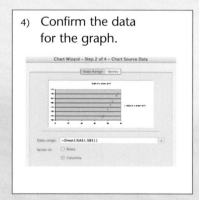

5) Choose graph options.

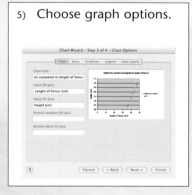

6) Choose the location of the graph.

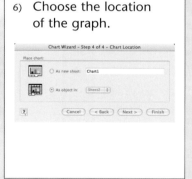

7) Draw the graph.

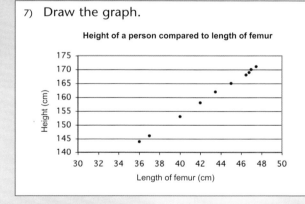

After drawing the graph you can modify different elements by double-clicking on the element to be changed: title, scale, legend, grid, type of graph, etc.

Below are different types of graphs you can create using a spreadsheet:

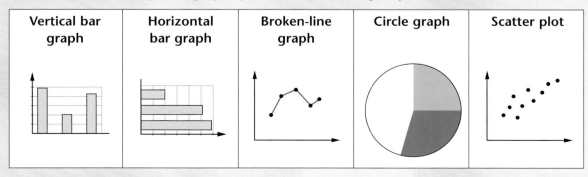

| Vertical bar graph | Horizontal bar graph | Broken-line graph | Circle graph | Scatter plot |

Dynamic geometry software

Dynamic geometry software allows you to draw and move objects in a workspace. The dynamic aspect of this type of software allows you to explore and verify geometric properties and to validate constructions.

The workspace and the tools

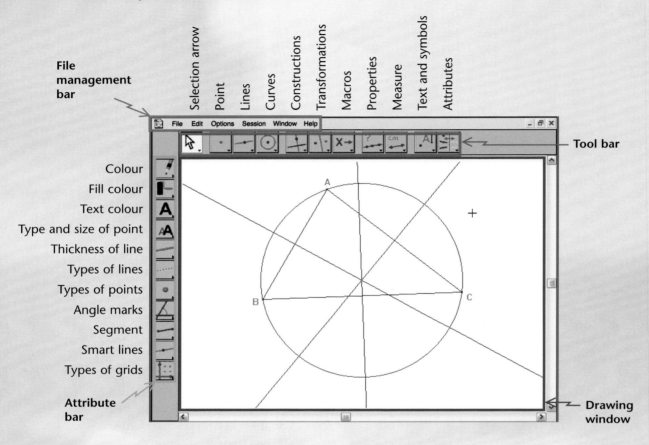

Cursors and their interpretations

+	Cursor used when moving in the drawing window.
☝	Cursor used when drawing an object.
What object?	Cursor used when there are several objects.
☝	Cursor used when tracing an object.
☜	Cursor used to indicate movement of an object is possible.
☝	Cursor used when working in the file management bar and in the tool bar.
☝	Cursor used when filling an object with a colour.
✐	Cursor used to change the attribute of the selected object.

Geometric explorations

1) A median separates a triangle into two other triangles. In order to explore the properties of these two triangles, perform the following construction. To verify that triangles ABD and ADC have the same area, calculate the area of each triangle. By moving the points A, B and C, notice that the areas of the two triangles are always the same.

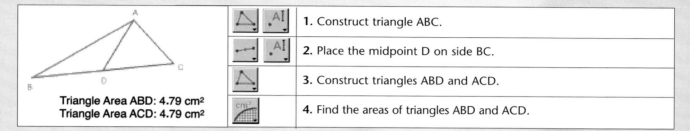

Triangle Area ABD: 4.79 cm²
Triangle Area ACD: 4.79 cm²

	1. Construct triangle ABC.
	2. Place the midpoint D on side BC.
	3. Construct triangles ABD and ACD.
	4. Find the areas of triangles ABD and ACD.

2) In order to determine the relation between the position of the midpoint of the hypotenuse in a right triangle and the three vertices of the triangle, perform the construction below. By moving points A, B, and C, note that the midpoint of the hypotenuse of a right triangle is equidistant from its three vertices.

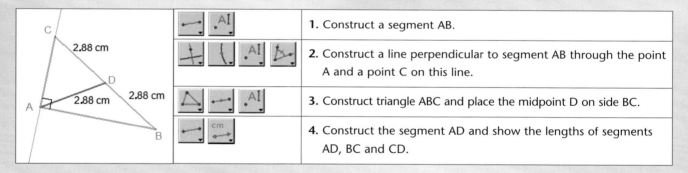

	1. Construct a segment AB.
	2. Construct a line perpendicular to segment AB through the point A and a point C on this line.
	3. Construct triangle ABC and place the midpoint D on side BC.
	4. Construct the segment AD and show the lengths of segments AD, BC and CD.

Graphical exploration

In order to discover the relation between the slopes of two perpendicular lines in the Cartesian plane, perform the construction below. By showing the product of the slopes and modifying the inclination of one of the lines, note a particular property of these slopes: the product of the slopes of these two perpendicular lines is -1.

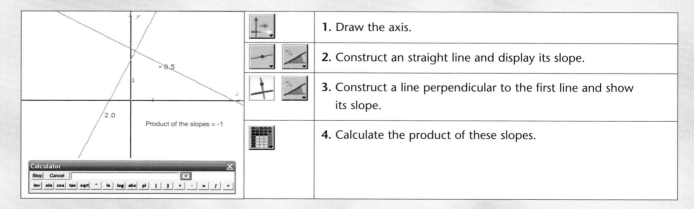

	1. Draw the axis.
	2. Construct an straight line and display its slope.
	3. Construct a line perpendicular to the first line and show its slope.
	4. Calculate the product of these slopes.

Notations and symbols

Notation & symbols	Meaning
{ }	Brace brackets, used to identify the elements in a set
$\mathbb{N}$	The set of Natural numbers
$\mathbb{Z}$	The set of Integers
$\mathbb{Q}$	The set of Rational numbers
$\mathbb{Q}'$	The set of Irrational numbers
$\mathbb{R}$	The set of Real numbers
$\cup$	The union of sets
$\cap$	The intersection of sets
Ω	Read "omega," it represents the sample space in a a random experiment
or { }	The empty set (or the null set)
=	...is equal to...
$\neq$	...is not equal to...
$\approx$	...is approximately equal to...
<	...is less than...
>	...is greater than...
$\leq$	...is less than or equal to...
$\geq$	...is greater than or equal to...
$[a, b]$	Interval, including a and b
$[a, b[$	Interval, including a but excluding b
$]a, b]$	Interval, excluding a but including b
$]a, b[$	Interval, excluding a and b
∞	Infinity
(a, b)	The ordered pair a and b
$f(x)$	Is read as f of x or the value (image) of the function f at x
Δx	Variation or growth in x

Notation & symbols	Meaning
( )	Parentheses show which operation to perform first
$-a$	The opposite of a
$\frac{1}{a}$ or a^{-1}	The reciprocal of a
a^2	The second power of a or a squared
a^3	The third power of a or a cubed
$\sqrt{a}$	The square root of a
$\sqrt[3]{a}$	The cube root of a
$|a|$	The absolute value of a
%	Percent
$a : b$	The ratio of a to b
π	Read "pi," it is approximately equal to 3.1416
$^\circ$	Degree, unit of angle measure
$\overline{AB}$	Segment AB
m $\overline{AB}$	Measure of segment AB
$\angle$	Angle
m $\angle$	The measure of an angle
$\overset{\frown}{AB}$	Arc AB
m $\overset{\frown}{AB}$	The measure of arc AB
//	… is parallel to…
$\perp$	… is perpendicular to …
⦜	Indicates a right angle in a plane figure
$\triangle$	Triangle
$\cong$	…is congruent to…
~	…is similar to…
$\overset{\wedge}{=}$	…corresponds to…
$P(E)$	The probability of event E
Med	The median of a distribution

Geometric statements

	Statement	Example
1.	If two lines are parallel to a third line, then they are all parallel to each other.	If $l_1 \parallel l_2$ and $l_2 \parallel l_3$, then $l_1 \parallel l_3$. 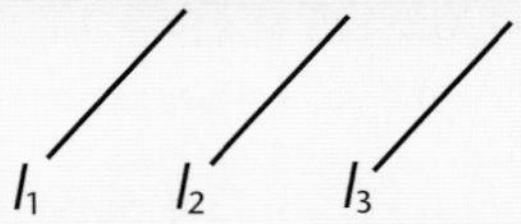
2.	If two lines are perpendicular to a third line, then the two lines are parallel to each other.	If $l_1 \perp l_3$ and $l_2 \perp l_3$, then $l_1 \parallel l_2$. 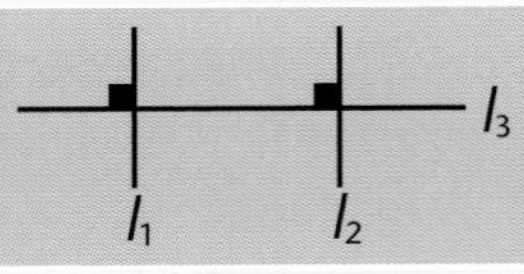
3.	If two lines are parallel, then every line perpendicular to one of these lines is perpendicular to the other.	If $l_1 \parallel l_2$ and $l_3 \perp l_2$, then $l_3 \perp l_1$. 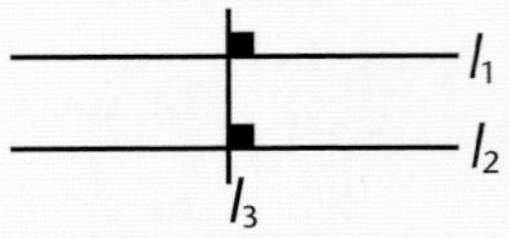
4.	If the exterior arms of two adjacent angles are collinear, then the angles are supplementary.	The points A, B and D are collinear. $\angle$ ABC & $\angle$ CBD are adjacent and supplementary. 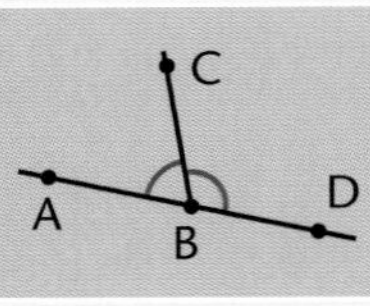
5.	If the exterior arms of two adjacent angles are perpendicular, then the angles are complementary.	$\overline{AB} \perp \overline{BD}$ $\angle$ ABC and $\angle$ CBD are adjacent and complementary. 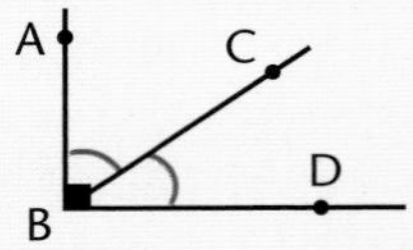
6.	Vertically opposite angles are congruent.	$\angle 1 \cong \angle 3$ $\angle 2 \cong \angle 4$ 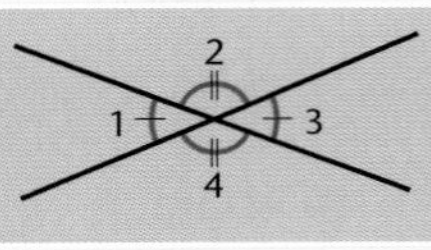
7.	If a transversal intersects two parallel lines, then the alternate interior, alternate exterior and corresponding angles are respectively congruent.	If $l_1 \parallel l_2$, then angles 1, 3, 5 and 7 are congruent as are angles 2, 4, 6 and 8. 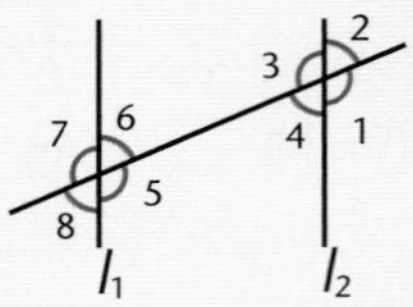
8.	If a transversal intersects two lines resulting in congruent corresponding angles (or alternate interior angles or alternate exterior angles), then those two lines are parallel.	In the figure for statement 7, if the angles 1, 3, 5 and 7 are congruent and the angles 2, 4, 6 and 8 are congruent, then $l_1 \parallel l_2$.
9.	If a transversal intersects two parallel lines, then the interior angles on the same side of the transversal are supplementary.	If $l_1 \parallel l_2$, then $m \angle 1 + m \angle 2 = 180°$ and $m \angle 3 + m \angle 4 = 180°$. 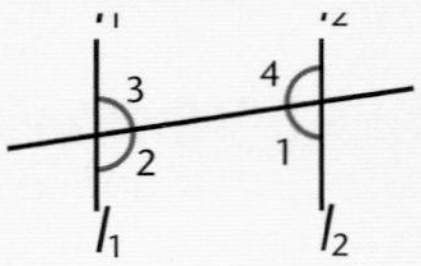

	Statement	Example
10.	The sum of the measures of the interior angles of a triangle is 180°.	$m \angle 1 + m \angle 2 + m \angle 3 = 180°$
11.	Corresponding elements of congruent plane or solid figures have the same measurements.	$\overline{AD} \cong \overline{A'D'}$, $\overline{CD} \cong \overline{C'D'}$, $\overline{BC} \cong \overline{B'C'}$, $\overline{AB} \cong \overline{A'B'}$ $\angle A \cong \angle A'$, $\angle B \cong \angle B'$, $\angle C \cong \angle C'$, $\angle D \cong \angle D'$
12.	In an isosceles triangle, the angles opposite the congruent sides are congruent.	In the isosceles triangle ABC: $\overline{AB} \cong \overline{AC}$ $\angle C \cong \angle B$
13.	The axis of symmetry of an isosceles triangle represents a median, a perpendicular bisector, an angle bisector and an altitude of the triangle.	Axis of symmetry of triangle ABC, Median from point A Perpendicular bisector of the side BC Bisector of angle A Altitude of the triangle
14.	The opposite sides of a parallelogram are congruent.	In the parallelogram ABCD: $\overline{AB} \cong \overline{CD}$ and $\overline{AD} \cong \overline{BC}$
15.	The diagonals of a parallelogram bisect each other.	In the parallelogram ABCD: $\overline{AE} \cong \overline{EC}$ and $\overline{DE} \cong \overline{EB}$
16.	The opposite angles of a parallelogram are congruent.	In the parallelogram ABCD: $\angle A \cong \angle C$ and $\angle B \cong \angle D$
17.	In a parallelogram, the sum of the measures of two consecutive angles is 180°.	In the parallelogram ABCD: $m \angle 1 + m \angle 2 = 180°$ $m \angle 2 + m \angle 3 = 180°$ $m \angle 3 + m \angle 4 = 180°$ $m \angle 4 + m \angle 1 = 180°$
18.	The diagonals of a rectangle are congruent.	In the rectangle ABCD: $\overline{AC} \cong \overline{BD}$
19.	The diagonals of a rhombus are perpendicular.	In the rhombus ABCD: $\overline{AC} \perp \overline{BD}$
20.	The measure of an exterior angle of a triangle is equal to the sum of the measures of the interior angles at the other two vertices.	$m \angle 3 = m \angle 1 + m \angle 2$

	Statement	Example
21.	In a triangle the longest side is opposite the largest angle.	In triangle ABC, the largest angle is A, therefore the longest side is BC. 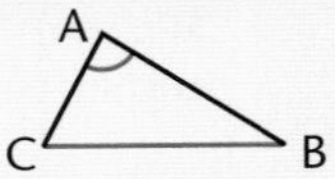
22.	In a triangle, the smallest angle is opposite the smallest side.	In triangle ABC, the smallest angle is B, therefore the smallest side is AC. 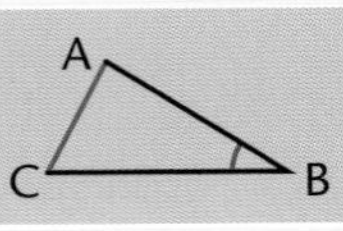
23.	The sum of the measures of two sides in a triangle is larger than the measure of the third side.	$2 + 5 > 4$ $2 + 4 > 5$ $4 + 5 > 2$ 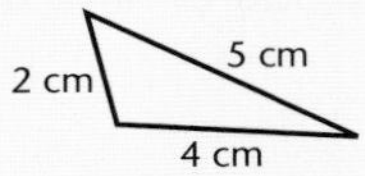
24.	The sum of the measures of the interior angles of a quadrilateral is 360°.	$m\angle 1 + m\angle 2 + m\angle 3 + m\angle 4 = 360°$ 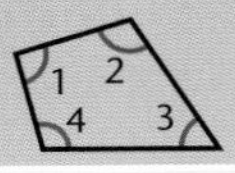
25.	The sum of the measures of the interior angles of a polygon with n sides is $n \times 180° - 360°$ or $(n-2) \times 180°$.	$n \times 180° - 360°$ or $(n-2) \times 180°$ 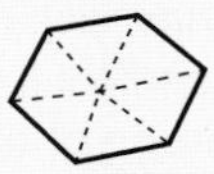 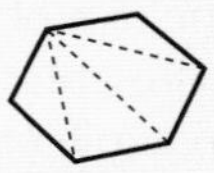
26.	The sum of the measures of the exterior angles (one at each vertex) of a convex polygon is 360°.	$m\angle 1 + m\angle 2 + m\angle 3 + m\angle 4 + m\angle 5 + m\angle 6 = 360°$ 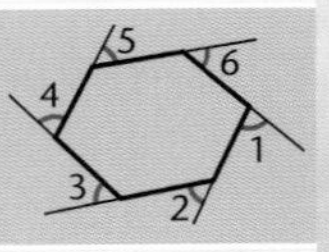
27.	The corresponding angles of similar plane figures or of similar solids are congruent and the measures of the corresponding sides are proportional.	The triangle ABC is similar to triangle A'B'C': $\angle A \cong \angle A'$ $\angle B \cong \angle B'$ $\angle C \cong \angle C'$ $\frac{m\overline{A'B'}}{m\overline{AB}} = \frac{m\overline{B'C'}}{m\overline{BC}} = \frac{m\overline{A'C'}}{m\overline{AC}}$ 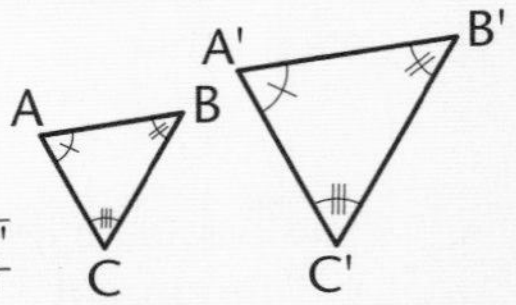
28.	In similar plane figures, the ratio of the areas is equal to the square of the ratio of similarity.	In the above figures, $\frac{m\overline{A'B'}}{m\overline{AB}} = \frac{m\overline{B'C'}}{m\overline{BC}} = \frac{m\overline{A'C'}}{m\overline{AC}} = k$ Ratio of similarity $\frac{\text{area of triangle A'B'C'}}{\text{area of triangle ABC}} = k^2$
29.	Three non-collinear points define one and only one circle.	There is only one circle which contains the points A, B and C. 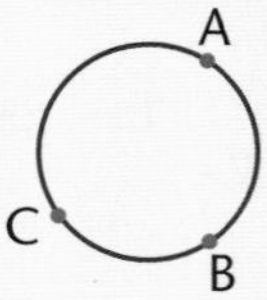
30.	The perpendicular bisectors of any chords in a circle intersect at the centre of the circle.	l_1 and l_2 are the perpendicular bisectors of the chords AB and CD. The point of intersection M of these perpendicular bisectors is the centre of the circle. 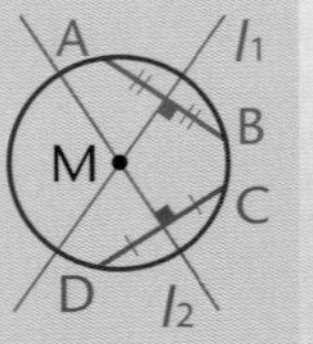

	Statement	Example
31.	All the diameters of a circle are congruent.	$\overline{AD}$, $\overline{BE}$ and $\overline{CF}$ are diameters of the circle with centre O. $\overline{AD} \cong \overline{BE} \cong \overline{CF}$
32.	In a circle, the measure of the radius is one-half the measure of the diameter.	$\overline{AB}$ is a diameter of the circle with centre O. m $\overline{OA} = \frac{1}{2}$ m $\overline{AB}$
33.	In a circle, the ratio of the circumference to the diameter is a constant represented by π.	$\frac{C}{d} = \pi$
34.	In a circle, a central angle has the same degree measure as the arc contained between its sides.	In the circle with centre O, m $\angle$ AOB = m $\overarc{AB}$ is stated in degrees.
35.	In a circle, the ratio of the measures of two central angles is equal to the ratio of the arcs intercepted by their sides.	$\frac{m\angle AOB}{m\angle COD} = \frac{m\ \overarc{AB}}{m\ \overarc{CD}}$
36.	In a circle, the ratio of the areas of two sectors is equal to the ratio of the measures of the angles at the centre of these sectors.	$\frac{\text{Area of the sector AOB}}{\text{Area of the sector COD}} = \frac{m\angle AOB}{m\angle COD}$

Glossary

A

Angles
Classification of angles according to their measure

Name	Measure	Representation
Zero	0°	
Acute	Between 0° & 90°	
Right	90°	
Obtuse	Between 90° & 180°	
Straight	180°	
Reflex	Between 180° & 360°	
Perigon	360°	

Alternate exterior, p. 204
Alternate interior, p. 204
Complementary, p. 77
Corresponding, p. 204
Supplementary, p. 77

Apothem of a regular polygon
Segment (or length of segment) from the centre of the regular polygon perpendicular to any of its sides. It is determined by the centre of the regular polygon and the midpoint of any side.

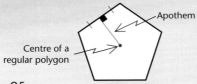

Centre of a regular polygon — Apothem

Arc cosine, p. 95

Arc sine, p. 95

Arc tangent, p. 95

Area
The measure of the surface of a figure. Area is expressed in square units.

Area of a circle

$$A_{circle} = \pi r^2$$

Area of a parallelogram

$$A_{parallelogram} = b \times h$$

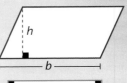

Area of a rectangle

$$A_{rectangle} = b \times h$$

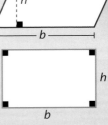

Area of a regular polygon

$$A_{regular\ polygon} = \frac{perimeter\ of\ polygon \times apothem}{2}$$

Area of a rhombus

$$A_{rhombus} = \frac{D \times d}{2}$$

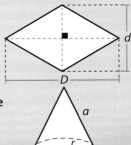

Area of a right circular cone

$$A_{right\ circular\ cone} = \pi r^2 + \pi r a$$

Area of a sector

$$\frac{\left(\begin{array}{c}\text{Measure of the central}\\\text{angle of a sector}\end{array}\right)}{360°} = \frac{sector\ area}{\pi r^2}$$

Area of a sphere

$$A_{sphere} = 4\pi r^2$$

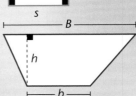

Area of a square

$$A_{square} = s \times s = s^2$$

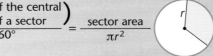

Area of a trapezoid

$$A_{trapezoid} = \frac{(B + b)h}{2}$$

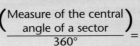
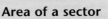

Area of a triangle

$$A_{triangle} = \frac{b \times h}{2}$$

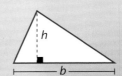

Arrangement, p. 141

Asymptote, p. 39

Capacity
Volume of a fluid which a solid can contain.

Cartesian plane
A plane formed by two graduated perpendicular lines.

Circumference
The perimeter of a circle. In a circle whose circumference is C, diameter is d and radius is r: $C = \pi d$ and $C = 2\pi r$.

Combination, p. 141

Coordinates of a point
Each of the two numbers used to describe the position of a point in a Cartesian plane.

Cosine of an angle, pp. 84, 95

Cube root
The inverse of the operation which consists of cubing a number is called finding the cube root. The symbol for this operation is $\sqrt[3]{\ }$.

E.g.   1) $\sqrt[3]{125} = 5$
        2) $\sqrt[3]{-8} = -2$

D

Degree of a monomial
The sum of the exponents of the monomial.
E.g.   1) The degree of the monomial 9 is 0.
        2) The degree of the monomial $-7xy$ is 2.
        3) The degree of the monomial $15a^2$ is 2.

Degree of a polynomial in one variable
The largest exponent of that variable in the polynomial.
E.g.   The degree of the polynomial $7x^3 - x^2 + 4$ is 3.

Dependent variable - see Variable.

Domain of a function, p. 7

E

Edge
Segment formed by the intersection of any two faces of a solid.

Equation
Mathematical statement of equality involving one or more variables.
E.g. $4x - 8 = 4$

Equation of a line
An equation which represents a relationship between two variables.

Equivalent equations
Equations having the same solution.
E.g. $2x = 10$ and $3x = 15$ are equivalent equations, because the solution of each is 5.

Events
Compatible, p. 132
Complementary, p. 132
Incompatible, p. 132

Expectation
Of gain, p. 164
Mathematical, p.164

Experiment
Random, p. 133
Random with several steps, p. 133
Random with order, p. 140
Random without order, p. 140

Exponentiation
Operation which consists of raising a base to an exponent.
E.g.   In 5^8, the base is 5 and the exponent is 8.

Extrema of a function, p. 7

F

Face
Plane or curved surface bound by edges.

Factorial, p. 140

Factoring
Writing an expression as a product of factors.
E.g.   The factorization of $6a^2 + 15a$ can be expressed as $3a(2a + 5)$.

Families of functions, p. 17

Function, p. 6
A relation between two variables in which each value of the independent variable is associated with at most one value of the dependent variable.

Direct variation function
A function in which a constant change in the independent variable results in a constant, non-zero change to the dependent variable. Its graph is an oblique line passing through the origin of the Cartesian plane.

First-degree polynomial function
A function whose rule can be written as a first-degree polynomial.
E.g. $f(x) = 7.1x + 195$

Inverse variation function
A function that represents an inversely proportional situation. The product of each ordered pair is a constant and non-zero. The graphical representation is a curve whose extremities gradually approach the axes but never touches them.

Partial variation function
A function in which a constant change in the independent variable results in a constant, non-zero change to the dependent variable. Its graph is an oblique line which does not pass through the origin of the Cartesian plane.

Periodic function, p. 53

Piecewise function, p. 53

Polynomial function
A function whose rule can be written as a polyminal.
E.g. $f(x) = 3x^2 + 7$

Quadratic function
Synonym for a second-degree polyminal function.

Second-degree polyminal function, pp. 28, 29

Sign of a function, p. 8

Step function, p. 53

Zero-degree polynomial function (constant function)
A function in which a constant change in the independent variable results in no change in the dependent variable. Its graph is a horizontal line that is parallel to the x-axis.
E.g. $f(x) = -5$

H

Hero's formula, p. 108

Height of a triangle (altitude)
Segment from one vertex of a triangle to the line containing the opposite side.

Hypotenuse
The side that is opposite the right angle in a right triangle. It is the longest side in a right triangle.

I

Independent variable - see Variable

Inequality
A mathematical statement which compares two numerical expressions with an inequality symbol (which may include variables).
E.g.   1) $4 < 4.2$
       2) $-10 \leq -5$

Integer
A number belonging to the set
$\mathbb{Z} = \{...,-2, -1, 0, 1, 2, 3, ...\}$.

Interval
A set of all the real numbers between two given numbers called the endpoints. Each endpoint can be either included or excluded in the interval.
E.g.   The interval of real numbers from -2 included to 9 excluded is [-2, 9[.

Irrational number
A number which cannot be expressed as a ratio of two integers, and whose decimal representation is non-periodic and non-terminating.

Laws of exponents

Law
Product of powers: $a^m \times a^n = a^{m+n}$
Quotient of powers: For $a \neq 0$ $\quad \dfrac{a^m}{a^n} = a^{m-a^n}$
Power of a product: $(ab)^m = a^m bc$
Power of a power: $(a^m)^n = a^{mn}$
Power of a quotient: $b \neq 0$: $\quad \left(\dfrac{a}{b}\right)^m = \dfrac{a^m}{bc}$

Legs (or arms) of a right triangle
The sides that form a right angle in a right triangle.

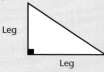

Leg

Leg

Like terms - see Terms.

Mathematical model, pp. 17, 18

Maximum of a function, p. 7

Median of a triangle
Segment determined by a vertex and the midpoint of the opposite side.
E.g. The segments
    AE, BF and CD
    are the medians
    of triangle ABC.

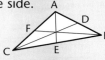

Metric relations (in a right triangle)

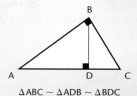

$\triangle ABC \sim \triangle ADB \sim \triangle BDC$

- In a right triangle, the length of a leg is the geometric mean between the length of its projection on the hypotenuse and the length of the hypotenuse.

$\dfrac{m\,\overline{AD}}{m\,\overline{AB}} = \dfrac{m\,\overline{AB}}{m\,\overline{AC}}$ or $(m\,\overline{AB})^2 = m\,\overline{AD} \times m\,\overline{AC}$

- In a right triangle, the length of the altitude drawn from the right angle is the geometric mean of the length of the two segments that determine the hypotenuse.

$\dfrac{m\,\overline{AD}}{m\,\overline{BD}} = \dfrac{m\,\overline{BD}}{m\,\overline{CD}}$ or $(m\,\overline{BD})^2 = m\,\overline{AD} \times m\,\overline{CD}$

- In a right triangle, the product of the length of the hypotenuse and its corresonding altitude is equal to the product of the length of the legs.

$m\,AC \times m\,BD = m\,AB \times m\,BC$

Minimum of a function, p. 7

Monomial
Algebraic expression formed by one number, one variable or a product of numbers and variables.
E.g. 9, $-5x^2$ and $4xy$ are monomials.

Natural number
Any number belonging to the set

$\mathbb{N} = \{0, 1, 2, 3, \dots\}.$

Numerical coefficient of a term
Numerical value multiplied by the variable or variables of a term.
E.g. In the algebraic expression
    $x + 6xy - 4.7y$, the numerical coefficients of the first, second and third terms are 1, 6 and -4.7, respectively.

Odds
Against, p. 153
For, p. 153

Origin of a Cartesian plane
The point of intersection of the two axes in a Cartesian plane. The coordinates of the origin are (0, 0).

Parabola, p. 28

Perimeter
The length of the boundary of a closed figure. It is expressed in units of length.

Period of a periodic function, p. 53

Permutation, p. 140

Perpendicular bisector
A perpendicular line passing through the midpoint of a segment. It is also an axis of symmetry for the segment.
E.g.

Polygon
A closed plane figure with three of more sides.

Polygons

Number of Sides	Name of Polygon
3	Triangle
4	Quadrilateral
5	Pentagon
6	Hexagon
7	Heptagon
8	Octagon
9	Nonagon
10	Decagon
11	Undecagon
12	Dodecagon

Regular polygon
A polygon where all sides are congruent and all angles are congruent.

Polyhedron
A solid determined by plane polygonal faces. E.g.

Polynomial
An algebraic expression containing one or more terms. E.g. $x^3 + 4x^2 - 18$

Prism
A polyhedron with two congruent parallel faces called "bases." The parallelograms defined by the corresponding sides of these bases are called the "lateral faces."

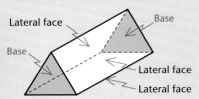

Probability
Of an event, p. 133
Experimental, p. 152
Subjective, p. 152
Theoretical, p. 152

Properties of a function, p. 7

Regular prism
A prism whose bases are regular polygons.
E.g. A regular heptagonal prism.

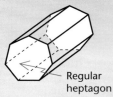

Right prism
A prism whose lateral faces are rectangles.
E.g. A right trapezoidal prism.

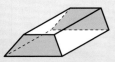

Proportion, p. 78

Pyramid
A polyhedron with one polygonal base, whose lateral faces are triangles with a common vertex called the "apex."
E.g. Octagonal pyramid.

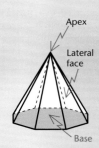

Regular pyramid
A pyramid whose base
is a regular polygon.
E.g. A regular hexagonal
pyramid.

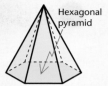
Hexagonal
pyramid

Right pyramid
A pyramid such that the segment
from the apex, perpendicular to
the base, intersects it at the
centre of the polygonal base.
E.g. A right rectangular pyramid.

Pythagorean theorem
In a right triangle, the sum of the squares of
the legs is equal to the square of the
hypotenuse.
E.g.

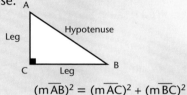

$$(m\overline{AB})^2 = (m\overline{AC})^2 + (m\overline{BC})^2$$

 Q

Quadrant
Each of the four regions determined by the
axis of a Cartesian plane. The quadrants
are numbered 1 to 4.

 R

Radius
A radius is a segment (or length of a segment)
which is determined by the centre of a circle
and any point on the circle.

Radius

Range, p. 7

Rate
A way of comparing two quantities or two
sizes expressed in different units and which
requires division.

Rate of change
In a relation between two variables, a
comparison between two corresponding
variations.

$$\text{Rate of change} = \frac{\left(\begin{array}{c}\text{variation of a}\\\text{dependent variable}\end{array}\right)}{\left(\begin{array}{c}\text{variation of an}\\\text{independent variable}\end{array}\right)}$$

Ratio
A way of comparing two quantities or two
sizes expressed in the same units and which
requires division.

Rational number
A number which can be written as the
quotient of two integers where the
denominator is not zero. Its decimal
representation can be terminating or non-
terminating and periodic.

Ratio of similarity
Ratio of corresponding segments resulting
from a dilatation.

Real number
A number belonging to the union of the set
of rational numbers and the set of irrational
numbers.

Relation
A relationship between two variables.

Removing a common factor
Writing an expression as a product of two
factors, one of which is common to the terms
of the original expression.
E.g. $8a^2 - 44a = 4a(2a - 11)$

Right circular cone
Solid made of two faces, a circle
or a sector. The circle is the base
and the sector forms the
lateral face.

Right circular cylinder
Solid made of three faces,
two congruent circles and a
rectangle. The circles form
the bases and the rectangle
forms the lateral face.

Rules for transforming equations
Rules that result in obtaining equivalent
equations. Solve an equation by respecting
the following:
• Adding or subtracting the same number on
  both sides of the equation.
• Multiplying or dividing both sides of the
  equation by a number other than 0.

Rules for transforming inequalities

Rules that result in obtaining equivalent inequalities.

- Adding or subtracting the same number on both sides of the inequality conserves the direction of the inequality.
- Multiplying or dividing both sides of the inequality by a positive number conserves the direction of the inequality.
- Multiplying or dividing both sides of the inequality by a negative number switches the direction of the inequality.

S

Scientific notation

A notation which facilitates the reading and writing of numbers which are very large or very small.

E.g.   1) $56\,000\,000 = 5.6 \times 10^7$
        2) $0.000\,000\,008 = 8 \times 10^{-9}$

Similar figures

Two figures are similar if and only if a dilation enlargement or reduction of one results in a figure congruent to the other.

Sine law, p. 108

Sine of an angle, pp. 84, 95

Slant height of a right circular cone

Segment (or length of a segment) defined by the apex and any point on edge of the base.
E.g.

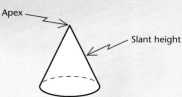

Slant height of a regular pyramid

Segment from the apex perpendicular to any side of the polygon forming the base of the pyramid. It corresponds to the altitude of a triangle which forms a lateral face.

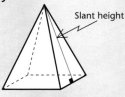

Solid

Portion of space bounded by a closed surface.
E.g.

Solving any triangle, pp. 84, 95, 108

Sphere

The set of all points in space at a given distance (radius) from a given point (centre).

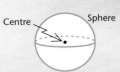

Square root

The inverse of the operation which consists of squaring a positive number is called finding the square root. The symbol for this operation is $\sqrt{\ }$.
E.g. The square root of 25, written $\sqrt{25}$, is 5.
    Note: $\sqrt{25}$ is called a "radical" and 25 is called the "radicand."

Surface area - see Area.

T

Tangent of an angle, pp. 84, 95

Terms

Coefficient of a term

The number preceding the variable(s) of an algebraic term.
E.g. In the algebraic expression $x + 6xy - 4.7y$, 1, 6 and 4.7 are, respectively, the coefficients of the first, second and third terms.

Like terms

Terms composed of constant terms or the same variables raised to the same exponents.
E.g. 1) $8ax^2$ and ax^2 are like terms.
     2) 8 and 17 are like terms.

Algebraic term

A term can be composed of one number or of a product of numbers and variables.
E.g. 9, x and $3xy^2$ are terms.

Trigonometric formula, p. 108

Trigonometric ratios, p. 84

U

Units of area

The square metre is the basic unit of area in the metric system (SI).

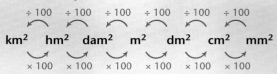

Units of capacity
The litre is the basic unit of capacity in the metric system (SI).

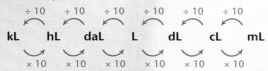

Units of length
The metre is the basic unit of length in the metric system (SI).

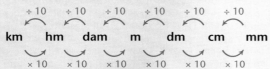

Units of volume
The cubic metre is the basic unit of volume in the metric system (SI).

$$\text{km}^3 \quad \text{hm}^3 \quad \text{dam}^3 \quad \text{m}^3 \quad \text{dm}^3 \quad \text{cm}^3 \quad \text{mm}^3$$
÷1000 ... ×1000

Variable
A symbol (generally a letter) which can take different values.

Variable
Dependent variable, p. 6
Independent variable, p. 6

Variation in a function, p. 7

Vertex of a solid
In geometry, a point common to at least two edges of a solid.

Volume
A measure of the space occupied by a solid, volume is expressed in cubic units.

Volume of a right circular cone
$$V_{cone} = \frac{(\text{area of the base}) \times (\text{height})}{3}$$

Volume of a right circular cylinder
$$V_{right\ circular\ cylinder} = (\text{area of the base}) \times (\text{height})$$

Volume of a right prism
$$V_{right\ prism} = (\text{area of the base}) \times (\text{height})$$

Volume of a pyramid
$$V_{pyramid} = \frac{(\text{area of the base}) \times (\text{height})}{3}$$

Volume of a sphere
$$V_{sphere} = \frac{4\pi r^3}{3}$$

x-axis (horizontal)
A scaled line which allows you to determine the x-value (abscissa) of any point in the Cartesian plane.

x-intercept (zero)
In a Cartesian plane, an x-intercept is the x-value (abscissa) of an intersection point of a curve with the x-axis.

x-value (abscissa)
The first coordinate of a point in the Cartesian plane.
E.g. The x-value (abscissa) of the point (5, -2) is 5.

y-axis (vertical)
A scaled line which allows you to determine the y-value (ordinate) of any point in the Cartesian plane.

y-intercept (initial value)
In a Cartesian plane, the y-intercept is the y-value (ordinate) of an intersection point of a curve with the y-axis

y-value (ordinate)
The second coordinate of a point in the Cartesian plane.
E.g. The y-value (ordinate) of the point (5, -2) is -2.

Photography Credits

T Top   **B** Bottom   **L** Left   **R** Right   **C** Center   **BG** Background

Cover

© Shutterstock

Vision 4

INTRO TL © Semjonow Juri/Shutterstock **INTRO TR** © Zphoto/Shutterstock **INTRO CL** © Hugo de Wolf/Shutterstock **INTRO CR** © Cary Kalscheuer/Shutterstock **4 TR** © Dusan Zidar/Shutterstock **5 TR** © Vling/Shutterstock **5 CR** © Thomas Mounsey/Shutterstock **10 BR** © Brian A Jackson/Shutterstock **11 TL** © Cartier Anne/Publiphoto **12 CL** © Pavlo Maydikov/iStockphoto **13 TR** © Hekimian Julien/Corbis Sygma **13 BC** © Galen Rowell/Corbis **15 BL** © Tari Faris/iStockphoto **22 BL** © Gunter Marx Photography/Corbis **23 C** © Photodisc/Getty **24 CR** © Maximilian STOCK LTD/Science Photo Library **25 TR** © United States Air Force **32 BL** © NASA/Science Photo Library **33 TL** © Anthony Hall/Shutterstock **33 CR** © United States Air Force **34 CR** © Mikael Damkier/Shutterstock **35 TR** © The London Art Archive/Alamy **37 TL** © Julián Rovagnati/Shutterstock **43 TR** © Debbie Oetgen/Shutterstock **45 TR** © Rafael Ramirez Lee/Shutterstock **46 TL** © Paul Prescott/Shutterstock **48 TR** © Sam Tinson/REX FEATURES/The Canadian Press **48 CR** © Sam Tinson/REX FEATURES/The Canadian Press **49 TR** © Alistair Scott/Shutterstock **49 BR** © Luchschen/Shutterstock **50 TR** © Ragnar Omarsson/Etsa/Corbis **51 CR** © Rafal Olkis/Shutterstock **54 CR** © Colman Lerner Gerardo/Shutterstock **57 TR** © Chrislofoto/Shutterstock **57 BL** © Desertsolitaire/Dreamstime.com **58 TR** © Gordon Ball LRPS/Shutterstock **58 CL** © Zoltan Pataki/Shutterstock **60 TL** 37068514 © 2008 Jupiter Images and its representatives **61 TR** © Science, Industry & Business Library/New York Public Library **61 BL** © TebNad/Shutterstock **62 TL** © Michael Mattner/iStockphoto **62 CL** © Ralph White/Corbis **62 CR** © Alexis Rosenfeld/Science Photo Library **62 C** © Alexis Rosenfeld/Science Photo Library **63 BL** © Royal Navy/epa/Corbis **64 TL** © Dr Jeremy Burgess/Science Photo Library **67 BL** © Clark Brennan/Alamy **69 BR** © The Gallery Collection/Corbis **70 BL** © Kelly-Mooney Photography/Corbis **71 TR** © Taiga/Shutterstock **72 TL** © Olga Kolos /Shutterstock **73 TR** © Papa Kay/Alamy **73 BC** 5267420 © 2008 Jupiter Images and its representatives

Vision 5

INTRO TL © Eric Gevaert/Shutterstock **INTRO TR** © Fatih Kocyildir/Shutterstock **INTRO CL** © Alfio Ferlito/Shutterstock **INTRO CR** © Jorge Salcedo/Shutterstock **76 BL** © NASA **79 CR** © Bulent Ince/iStockphoto **81 BL** © Carson Ganci/Design Pics/Corbis **87 CB** © Lijuan Guo/Shutterstock **89_TC** © Nathalie Ricard **90 TR** 8895989 © 2009 Jupiter Images and its representatives **90 BL** © US Department of Defense/Science Photo Library **91 BC** © Xiver/Shutterstock **97 CL** © Olivier Blondeau/iStockphoto **98 TR** © Roger Ressmeyer/Corbis **98 BC** © Gina Sanders/Shutterstock **99 BR** © Anton Foltin/Shutterstock **100 BC** © US Library of Congress/Science Photo Library **101 BR** © The London Art Archive/Alamy **105 TC** Royalty free photo **111 CR** © Mark Sykes/Alamy **114 CL** © Bettmann/Corbis **114 CR** 5268920 © 2009 Jupiter Images and its representatives **116 TL** © Douglas Peebles/Corbis **116 BC** © Kin Cheung/Reuters/Corbis **120 CL** © Fotohunter/Shutterstock **120 BR** © Mircea BEZERGHEANU/Shutterstock **122 TR** © Marc de Oliveira/iStockphoto **123 BC** © Stephen Oliver/Dorling Kindersley **124 CL** © Corbis Super RF/Alamy **124 BR** © Robert Trail/Dennis Milon/Science Photo Library **125 TR** © Rosanne Dubé **125 BL** © Rosanne Dubé **127 TR** © Tom Grundy/Shutterstock

Vision 6

INTRO TL © Joseph Gareri/Shutterstock **INTRO TR** © Dave Allen Photography/Shutterstock **INTRO CL** © Sebastian Kaulitzki/Shutterstock **INTRO CR** © Andrew Chin/Shutterstock **130 TR** © Tatyana Chernyak/iStockphoto **130 BR** © Photodisc **131 BL** © Parc du Mont-Comi **134 TR** © Michelle Milano/Shutterstock **134 BR** © Don Bayley/iStockphoto **135 TR** © Lyle Gregg/iStockphoto **135 BR** © Kenneth Cheung/Shutterstock **136 CL** © Beaux Arts/Alamy **137 TR** © Alvin Teo/iStockphoto **137 BL** © Corbis **138 BL** © Stephen Coburn/Shutterstock **142 TR** © J-L Charmet/Science Photo Library **143 CL** © Pali Rao/iStockphoto **144 CR** © Navarro Raphael/Shutterstock **144 BR** © Yury Kosourov/Shutterstock **145 CR** © AYAKOVLEVdotCOM/Shutterstock **145 CB_**© Shutterstock **146 C** © Olivier Berg/epa/Corbis **147 TR** © Francesco Abrignani/Shutterstock **147 CR** © K Chelette/Shutterstock **149 BC** © Mcmaster Studio/Alamy **154 TR** © Donna Coleman/iStockphoto **155 TR** © Monkey Business Images/Shutterstock **155 BR** © Vysokova Ekaterina/Shutterstock **156 CR** © Stephen Coburn/Shutterstock **158 BR** © Jo-Hanna Wienert/iStockphoto **159 TL** © Photodisc **159 TR** © KennStilger47/Shutterstock **159 CR** © Peter Casolino/Alamy **160 BL** © Boris Stroujko /Shutterstock **160 BC** © Quang Ho /Shutterstock **160 BR** © Lucian Coman /Shutterstock **161 TR** © iStockphoto **165 BR** © Hannah Eckman/Shutterstock **166 BR** © José Luis Gutiérrez/iStockphoto **168 CR** © Articular/Shutterstock **169 TR** © Clifford Skarstedt/The Canadian Press **169 BR** © Phil Talbot/Alamy **170 TL** 37009379 © 2008 Jupiter Images and its representatives **171 CL** © Bleex/iStockphoto **172 TL** © Zsolt Nyulaszi/Shutterstock **172 BL** © Pali Rao/iStockphoto **172 BR** © Lisa F. Young/Shutterstock **174 BR** © Christopher Henke/Shutterstock **175 CR** © Studio Araminta/Shutterstock **176 CL** © Coby Burns/ZUMA/Keystone Press **176 CB** © Brandon Laufenberg/iStockphoto **177 CL** © Tammy Bryngelson/iStockphoto **177 BR** © Shutterstock **178 TL** © ALBERT GEA/Reuters/Corbis **178 CR** © Hisom Silviu/Shutterstock **180 BR** Marc Wathieu, Raymond Queneau, cent mille milliards de poèmes © Éditions Gallimard **181 BL** © Chris Bott /Splash News/Keystone Press **182 CR** © Erzetic/Shutterstock

LES

184 TR © Kristian Sekulic/iStockphoto **184 CL** © Cindy Hughes/Shutterstock **184 CB** © prism_68/Shutterstock **185 TR** © egd/Shutterstock **185 CR** © Kristian Sekulic/Shutterstock **186 TR** © Dimitrije Paunovic/Shutterstock **187 TR** © Stephen Coburn/Shutterstock **187 BR** © Terekhov Igor/Shutterstock **188 BR** © The Canadian Press **189 TR** © Tina Spruce/iStockphoto **190 BR** © Marek Pawluczuk/Shutterstock **191 BL** © Gideon Mendel/Corbis **192 TR** © Michael Freeman/Corbis **193 TR** © Jeff Schultes/Shutterstock